TAMING THE SKIES

TAMING THE SKIES

A Celebration of Canadian Flight

Peter Pigott

A HOUNSLOW BOOK
A MEMBER OF THE DUNDURN GROUP
TORONTO · OXFORD

Publisher: Anthony Hawke
Copy-Editor: Andrea Pruss
Design: Jennifer Scott
Printer: Friesens

National Library of Canada Cataloguing in Publication Data

Pigott, Peter
Taming the skies : a celebration of Canadian flight / Peter Pigott.

ISBN 1-55002-469-8

1. Aeronautics--Canada--History. I. Title.

TL523.P528 2003   629.13'0971   C2003-902955-7

1   2   3   4   5   07   06   05   04   03

We acknowledge the support of the **Canada Council for the Arts** and the **Ontario Arts Council** for our publishing program. We also acknowledge the financial support of the **Government of Canada** through the **Book Publishing Industry Development Program** and **The Association for the Export of Canadian Books**, and the **Government of Ontario** through the **Ontario Book Publishers Tax Credit** program, and the **Ontario Media Development Corporation's Ontario Book Initiative**.

J. Kirk Howard, President

Printed and bound in Canada.✪
Printed on recycled paper.
www.dundurn.com

Dundurn Press
8 Market Street
Suite 200
Toronto, Ontario, Canada
M5E 1M6

Dundurn Press
73 Lime Walk
Headington, Oxford,
England
OX3 7AD

Dundurn Press
2250 Military Road
Tonawanda NY
U.S.A. 14150

For my mother-in-law,
Hazel O'Neill

ACKNOWLEDGEMENTS

As with all my other books, *Taming the Skies* would not have been possible without the help of many people. I list them in no particular order, each knowing that their generosity of time and spirit helped make it feasible.

They are Janet Lacroix at the Canadian Forces Joint Imagery Centre, Ottawa; Airman of the Year Master Warrant Officer Normand Marion, 16 Wing — I recommend the book he edited, *Camp Borden: Birthplace of the RCAF*; Captain Jeff Manney, 19 Wing; 2nd Lieutenant Natalie Oscien, 12 Wing; Captain Jason Proulx, also of 12 Wing; and Captain M.R. Perrault, 1 Wing.

My thanks also to Clive Prothero-Brooks, curator of the Royal Canadian Artillery Museum, CFB Shilo, Manitoba; to Robert Noorduyn, who edited my essay on his father's masterpiece; to Air Canada Captain Allan Snowie and his former naval classmate Andre Gignac for their memories and photos on the Cosmo; to Jerry Vernon for his research on the Mustang; to Air Canada Captains Robert Giguere and Hugh Campbell for the description of flying a DC-9 into Rockliffe; yet again to the family of bushpilot Art Schade for the use of his archives; to Scott Knox for his photos and history of the Halifax recovery and restoration; to Neil Aird and Rich Hulina for use of their photos; and finally, to my own father-in-law, George O'Neill, who shared with me his memories of what it was like to be twenty years old and part of a Sunderland crew.

To the ladies who brightened my day as I wrote this, at the Lester B. Pearson Library: Natalie Bisson, Jennifer Abou-Kheir, and Shana Glastonbury. And at home, as always, my wife and daughters.

TABLE OF CONTENTS

PREFACE
Fifty Aircraft in Context

When my last work, *Wings Across Canada: An Illustrated History of Canadian Aviation*, came out, one reviewer took exception to the title. He questioned how a collection of aircraft photographs could be called a history. As this work is similarly titled, an explanation is due.

In school, the history that I was taught consisted of the deeds of great men rather than the confluence of social and economic movements. In its name, I memorized an endless litany of kings, queens, and prime ministers — and their tombstone data. It was all character(s), and no context. But like most dichotomies, the split between character and context is a false one. For character gives context an essence, context is what brings character to life.

So it is with the fifty aircraft in this book. The Catalina flying boat would not be remembered as a great aircraft had it not been for its endurance, which allowed it to find German submarines and Japanese fleets. It served many Canadian heroes well — David Hornell and Len Birchall being the best known. But what made it truly magnificent was that it became representative, through its wide-ranging use, of a whole generation of pre-war aircraft — Sunderlands, Defiants, and Stranraers, all conceived in the 1930s but adapted to the conflict. Similarly, the de Havilland Comet was lethally flawed and uneconomical to operate, yet it was a brilliant and daring work of aeronautical engineering, and among the first customers were the Royal Canadian Air Force and Canadian Pacific Airlines. The Silver Dart was quickly outmoded and brought J.A.D. McCurdy to grief at air meets; nevertheless, as the first aircraft to fly in this country, it is an icon as much as the Wright brothers' Flyer is to the United States. The Mustang, the Mosquito, and the Halifax all had their faults; however, their use in daring raids and strategic bombing helped change the course of the Second

World War. Between the wars, the Vedette and the Norseman opened up a Canada far from train tracks and cities. In the 1960s, many Canadians took their first jet flight in the Douglas DC-9, and the Avro Jetliner and Canadair Dynavert show that we take second place to no one when it comes to aeronautical ingenuity. As for the de Havilland Beaver, to quote the artist Neil Aird, it is as Canadian as the ubiquitous red or green canoe on the lake, albeit creating a little more noise when in motion. In short, every aircraft in this book — like the kings and prime ministers I memorized — should be seen in the context in which it served, and that is why each of their stories is so fascinating.

True, some aircraft were more memorable than others. Aviation historians apart, who can put a shape to the Airspeed Oxford or the S-51 helicopter? The RCAF was quick to abandon their Northrop Deltas in 1941, and the forty-year-old Sea King is a haunting embarrassment that the government of Prime Minister Jean Chrétien would rather forget. But loved or not, each aircraft forms part of our country's story, fitting into the Canada that we have inherited. With the demise of many went part of our history. The Auster was the last of the artillery spotters, a profession that can be traced to the observation balloons of the American Civil War. The Vickers Vanguard was Air Canada's final mainline propeller aircraft, and the Boeing 767 was the last to wear the Canadian Airlines goose, a symbol that can be traced back to James Richardson and Canadian Airways. Thus to concede that history, whether it be kings or aircraft, is basically a collection of dates and photographs is not to render it irrelevant. For seen in the context of their times, such an accumulation as this allows us to connect the dots more easily and make sense of our own world.

INTRODUCTION

Our yearning to fly possessed us almost as soon as we began to walk upright. We must have watched the birds flying free and longed to escape this earthbound existence with them. We must have studied the heavens and yearned to explore their mysteries. But flying belonged to the gods alone, and in religious tales — including Judaism, Islam, and Christianity — only those that they favoured, the prophets and holy men, were allowed to conquer gravity. Ordinary people had to content themselves with inventing stories of flying dragons, winged horses, and magic carpets. But religion and folklore were not enough. If the story of Icarus is anything to go by, individuals of every race and era seem to have strapped feather-covered wings on themselves and jumped off towers. The wiser ones like Leonardo Da Vinci and Jules Verne designed fanciful machines that promised flight — if the necessary technology could catch up with the dream.

It took until December 17, 1903, at precisely 10:35 A.M. for humanity to unlock the mysteries of flight. Orville and Wilbur Wright were meticulous and secretive and attempted to monopolize the technology, but as one of the witnesses to that historic first flight summed it up: "Damned if they ain't flew!" To get there the two Ohio bicycle makers climbed on shoulders of giants like Octave Chanute, George Cayley, Otto Lilienthal, and Louis Mouillard. The twelve-second flight covered a distance of barely 120 feet but it fulfilled one of our oldest dreams. Perhaps the greatest legacy of all those pioneers was that they made the impossible happen. When it came to Canada, aviation captured the imagination of visionaries like J.A.D. McCurdy, W.R. Turnbull, Lawrence Lesh, and F.W. Baldwin, but not the closed minds of politicians in Ottawa.

A century later, aircraft have overturned our notions of time and space, of warfare and commerce, and the technology has gone from the Wright Flyer and the Silver Dart to the Global Express and the Stealth bomber. The twentieth century belongs to those who had the daring and foresight to create these machines, from experiments on North Carolina sands and frozen Nova Scotia lakes to high-performance fighters that have taken us to the far reaches of the atmosphere. For historic aircraft, whether designed for barnstorming, bombing, or global commuting, are the summation of the visions of their creators. The success of each is another of our dreams realized ... and that is where our story begins.

Ottawa, April 2003.

THE SILVER DART

It couldn't have been more Canadian, taking place as it did in midwinter and on the ice — on a frozen Nova Scotia lake in February 1909, to be exact. Orville and Wilbur Wright, who had flown the first controlled flying machine on the North Carolina sands six years before, had done so in great secrecy, their only eyewitnesses being the lifeguards at the beach station. In contrast, this was to be a very communal effort, with a crowd of neighbours and local women and children. It was a school holiday, young men on skates pulling at the machine, a horse-drawn sleigh leading the procession. Whatever lay in the future, the birth of aviation in Canada was a joyous occasion, as all such events should be.

It was nine years into the new century; the Victorian Age had been firmly laid to rest with the old Queen years before. There were now eight thousand automobiles in Toronto alone. The Boers and Boxers had had the temerity to take on the mighty British Empire, and women in Winnipeg were agitating for the vote. There were those Canadians who had read Sigmund Freud's newly published *The Interpretation of Dreams*, its contents, in which Freud had theorized that the wish to fly was connected with sexual accomplishment, were only slightly more bizarre than the stories that in Europe and the United States aircraft were actually flying.

As far as flight went, Canadians were familiar with balloons at agricultural fairs, but those fortunate enough to live around Bras d'Or Lake in Cape Breton, Nova Scotia, were somewhat more scientifically informed of the principles of flight. Alexander Graham Bell (yes, the inventor of the telephone) had his summer home Beinn Bhreagh, or "Beautiful Mountain," there, and it was where he pursued his aerodrome experiments of launching large tetrahedral kites that seemed devoid of practical purpose.

Baddeck No. 2 at Bentick Farm, Big Baddeck, with F.W. Baldwin in aviator's seat.

It was his friendship with Samuel Langley of the Smithsonian Institute in Washington that got Bell into designing large box kites at his Cape Breton home. In 1907, his wife, Mabel, suggested that he recruit younger talent to form an association with the purpose of carrying on experiments related to ... constructing a successful aerodrome. She also said that she would be prepared to finance such an enterprise. Bell's status and their own aeronautical curiosity attracted four young men to the Aerial Experiment Association. They were J.A. Douglas McCurdy, the son of Bell's secretary and a University of Toronto engineering student, his university colleague Frederick Casey Baldwin, American motorcycle racer Glenn Hammond Curtiss, and U.S. Army volunteer Lieutenant Thomas Selfridge. From the outset it must have been a clash in eras: on one side was the patriarch Bell, the woolly-minded eccentric who insisted on calling flying machines aerodromes; on the other were the steely purposed men of the new century, McCurdy, Baldwin, Selfridge, and above all Curtiss. Like the Wrights, Curtiss was also a bicycle builder and aircraft tinkerer and would be granted the first American pilot's licence. He was destined to play a crucial role in Canadian aviation, through his subsequent aircraft the JN-4 and the HS2L.

The dual purpose of the AEA was to continue to construct Bell's tetrahedral kites and to allow each of the young men to experiment with building and flying a machine of his own. The first effort was one of Bell's kites, Cygnet I. Towed behind a tugboat on Bras d'Or Lake on December 6, 1907, it was guided hang-glider style by Selfridge, achieving momentary liftoff before crashing. The second, Selfridge's Red Wing, would be more successfully piloted by Casey Baldwin at Hammondsport, New York, on March 12,1908. Baldwin became the first Canadian to fly. Baldwin's own design, White Wing, performed as well on May 18, but by far the most successful would be Curtiss's June Bug, which, on July 4, flew a full mile.

In both these last aircraft, a French invention called ailerons, movable surfaces at the edge of wings to improve lateral stability, were used for the first time in North America. In doing so, the AEA earned the wrath of Orville and Wilbur Wright, who, since 1905, had spent their time feuding with other aviation pioneers. The brothers held that by inventing controlled flight, especially with the device of wing warping for lateral control, all other aircraft designers owed them patent fees. That the AEA had achieved more in six months than he and his brother had in six years, Orville thought, could only be through illegal means. In truth, the Wrights were losing their lead in aviation to European and North American aviators and had cause to worry.

But the AEA pressed ahead, putting the June Bug on pontoons, renaming it the Loon, and continuing to perfect their designs through the summer. They lost Selfridge, who became the first person in history to be killed in an aircraft when on September 17, the Wright Flyer that he was a passenger in crashed at Fort Myer, Virginia. Bell, McCurdy, and Baldwin attended the funeral and, seeing the wrecked Flyer, surmised correctly that wing warping had been the cause of the crash. By the fall, only Douglas McCurdy's design had yet to be flown, and with the new year approaching, Bell reminded his young men that he had two kites, the Cygnet II and Oinos, to be completed, and he recalled them to Beinn Breagh.

Curtiss, chafing at the time wasted on Bell's quixotic kites, was preparing to leave the AEA, and he and McCurdy ignored the summons to Baddeck to work on the AEA's fourth aircraft. Incorporating much from the June Bug, the pair built the Silver Dart, so named because of the fabric used. It was 32 feet long, spanned 49 feet, and was 10 feet high. The silver wing area — the colour was chosen for the photographs — was 420 square feet. It was powered by the latest Curtiss engine, a water-cooled, 50-horsepower, eight-cylinder machine that drove a twin-bladed propeller. Not only did it have ailerons, for the first time there were seats for a pilot and passenger in tandem. On December 6, McCurdy would take the Silver Dart up for a distance of six hundred feet, then make ten similar flights. Perhaps guessing Curtiss's intentions, Bell asked that the Silver Dart be sent immediately to Baddeck. It arrived on February 6, 1909, to be assembled for its first flight in Canada.

By the time Curtiss and McCurdy returned to the Bell home, the ice on Bras d'Or Lake was judged firm enough for flying. On February 22, 1909, Mr. and Mrs. Bell, Curtiss and McCurdy (Baldwin was delivering a speech in Toronto), and half the townspeople of Baddeck had gathered on the frozen lake surface. Unlike the Wrights, Bell

Curtiss Loon hydroplane built by the Aerial Experiment Association, Hammondsport, New York, November 5, 1908.

Courtesy of the National Archives of Canada

Silver Dart aircraft of the Aerial Experimental Association, near Baddeck, Nova Scotia.

Courtesy of the National Archives of Canada

made a point of inviting everyone to his demonstrations. If the Wrights shunned publicity, the AEA sought it — to their advantage. The giant motorized kite Cygnet II (powered by Curtiss's engine from the Silver Dart) was brought out onto the ice. As everyone apart from Bell expected, it refused to become airborne, and in the effort, the propeller fell off. The next day, February 23, the engine was taken off the Cygnet and put back on the Silver Dart, along with a propeller from an iceboat. Around 1:00 P.M., its wings shimmering, the appropriately named aircraft was wheeled out of the Kite House and then dragged onto the ice by a ground crew in skates. The audience of 147 locals (all their names documented for posterity) included school children, who had been given the day off, and women, some of whom had sewn the fabric onto the wings. All must have watched the proceedings with mixed feelings. Although the old man in the horse-drawn sleigh with his wife was a world-famous celebrity, everyone knew that defying gravity was an impossibility. But en masse, they followed the craft and its crew out onto the ice. Whatever happened, this was going to be a homegrown spectacle, for it was local boy Douglas McCurdy who was operating the controls.

McCurdy took the Silver Dart up to some thirty feet, flying for about half a mile at about forty miles an hour, expertly landing it on the ice once more. Although it was over before the audience had an opportunity to appreciate it (Bell, who knew a thing or two about patents, did make sure that a photograph was taken), the ninety-second flight was history. Recognized by the Royal Aero Club, McCurdy had become the first British subject to fly a heavier-than-air machine anywhere in the British Empire.

On February 24, he guided the Silver Dart on a circuitous course of four and a half miles, but on landing its right wing struck the ice and a wheel collapsed. It was soon repaired, and by spring McCurdy was flying a full twenty miles in a closed circuit. Curtiss had by now left the AEA to pursue his own interests, and on March 31, the remaining three members gathered for a final time in the drawing room of Beinn Bhreagh. Baldwin moved that the association be dissolved, with McCurdy seconding.

The next month, Bell, Baldwin, and McCurdy went into business to begin Canada's first aircraft manufacturing company, the Canadian Aerodrome Company. With an American-built 42-horsepower Kirkham engine, the Baddeck 1 was the first aircraft to be completely built in Canada. Baldwin and McCurdy realized that the only customers for flying machines were the military, and that summer, the two young men took both aircraft to Ottawa for a demonstration. The politicians were out of town for the summer and the military was unenthusiastic — what use was such a machine on the battlefield? But in July, with their aircraft in crates, McCurdy and Baldwin arrived at Petawawa Army Camp outside the capital. The Department of National Defence allotted the sum of five dollars for the preparations, making it the first government expenditure on aviation in Canadian history. When the craft was assembled, on August 2, in bright sunshine, the officers, press, and public gathered to watch — the news that Bleriot had conquered the English Channel the week before perhaps providing the impetus to do so. McCurdy swung the propeller and climbed into the Silver Dart. He flew about half a mile at a height of ten feet before executing a faultless landing on the parade ground. On the next flight he took Baldwin up as a passenger, the first time this had happened in Canada. On the third, it was the turn of the foreman from the army workshop. By now the sun was low and, perhaps tired and overconfident, McCurdy made a fourth flight. This time he was blinded by the sun and hit a grassy hillock. The Silver Dart crashed onto its starboard wing, irretrievably damaged. Baddeck 1 fared little better. The audience was relieved to know that their beliefs could remain unshaken. They were almost beginning to think that aircraft had a future....

Dejected, McCurdy and Baldwin returned to Baddeck and to Bell's tetrahedral dreams. Fortune seems to have deserted McCurdy, as on June 24, 1910, he crashed Baddeck 2 at the Lakeside meet in Montreal. After other mishaps at the air meet at Donlands Farm in Todmorden, near Toronto, like many Canadians before and since, McCurdy would leave the country to seek his fortune below the border, working for former AEA member Glenn Curtiss.

Legend has it that the Silver Dart engine would be recycled to power a motorboat at Bras d'Or Lake. Fifty years later, on February 23, 1959, the first aircraft in Canada was replicated and flown once more at Baddeck on the anniversary of that first flight. With Bell, Baldwin, and Curtiss long since dead, only J.A.D. McCurdy, now the elder statesman of Canadian aviation, was on hand to witness the event and greet the pilot. Today, the Alexander Graham Bell National Historic Site on the east edge of the village of Baddeck commemorates Bell and the AEA. In private ownership, Beinn Bhreagh is maintained by Bell's descendants, and many of the historic outbuildings like the Kite House are still there. The anniversary replica of the Silver Dart is displayed at the National Aviation Museum, Ottawa. The original showed the way for every aircraft in Canada to follow.

SOPWITH CAMEL

To be a fighter pilot in the First World War required a natural affinity with flying machines. Young men who had grown up with horses treated their aircraft — be they Spads or Fokkers — as chargers in medieval warfare. They loved or hated, respected or despised them. To those who became aces and those who survived to the Armistice, the affinity had to be instinctive, as William Barker's was with the Sopwith Camel.

Herbert Smith's success with the Sopwith Pup in 1916 led him to redesign the aircraft into two other famous machines — the Triplane and the Camel. The Pup was the perfect fighting machine and a masterpiece to its pilots, and both the Admiralty and Royal Flying Corps (RFC) snapped it up in large numbers. Smith then drew up the design for the Camel, and with Sopwith and the Admiralty equally enthusiastic about it, on December 22, 1916, the decision was made to produce it. But compared with the Pup, the aptly named Camel proved to be a beast of a different temperament.

Initially fitted with a 110-horsepower Clerget 9Z rotary engine, it was test-flown at Brooklands on February 26, 1917, before being mass-produced. The Pup, like the Fokker DIII, had been armed with a single synchronized machine gun, which was why both were so manoeuvrable and delightful to fly. When higher powered engines became available, the next generation of fighters was armed with twin machine guns, able to deliver a lethal burst at an opponent and allow the pilot to escape before the enemy could retaliate. Typical of this new breed was the Camel. Thought under-powered, when a Clerget 130-horsepower was substituted for the 110-horsepower, it allowed for twin synchronized Vickers to be carried; the breeches were enclosed in an aluminum hump that earned the aircraft its name. The

Sopwith Camel.

Courtesy of the DND

aircraft were powered with a variety of other engines: the Bentley BR 1, Gnome Monosoupape, and Le Rhone 9J. Production began for the Naval and RFC squadrons, the first Camels arriving at the Front by June. By the end of 1917, well over one thousand Camels had been produced by Sopwith and others as shipboard fighters, armoured trench fighters, and, with a ceiling of nineteen thousand feet, a way for the Home Defence squadrons to get at the zeppelins.

In contrast with the Pup, the Camel was unpopular with its pilots. Like its namesake, it was unforgiving and unpredictable. Novices discovered that it was very sensitive to elevator controls and, because of its heavy rotary engine and short fuselage, too fast on right-hand turns. Yet dangerous as it was to fly, it was used to good purpose by many Allied air aces, especially William George Barker.

From Dauphin, Manitoba, Barker had escaped the trenches to be a Lewis gunner in the outdated Be2d over the Battle of the Somme. Getting his wings in January 1917, he transferred to 28 Squadron, where he first flew the Camel. By October of that year his score stood at five enemy aircraft and he was posted to the Italian front. Reeling under the Austrian onslaught, the Italians asked for RFC squadrons to boost their troops' morale. On November 29, 1917, in his cherished Camel B6313, Barker shot down his first Austrian aircraft, and after subsequent victories, he earned the nickname the artist with a pair of Vickers. Posted to 66 Squadron, as his score grew he had white flashes adorn the struts of B6313, keeping tally of his kills. Between April 17 and July 13, 1918, as the Battle of Piave raged below, sixteen of the enemy fell to his guns. In July, at the age of twenty-four, he was promoted to major and decorated with the Distinguished Service Order, Military Cross, Croix de Guerre, and Italian Valore Militare.

He was also given command of 139 Squadron. As they flew Bristol Fighters, Barker was forbidden to take the white-striped B6313 with him. He had it sent back to the aircraft depot and then returned to 139 on temporary

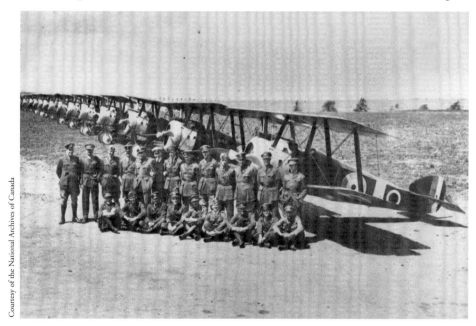

attachment so that he could continue to fly it. With the Italian front now secure, he was returned to Britain to teach new recruits aerial warfare. Soon tiring of that, he got himself transferred once more to the Front, to 201 Squadron, also equipped with Camels.

Squadron Commander Raymond Collishaw and pilots with Sopwith F.1 Camel aircraft of No. 203 Squadron, Allonville, France, July 1918.

Courtesy of the National Archives of Canada

25

His score now at forty-six enemy aircraft, Barker became Canada's most decorated soldier of both wars. But it was in a Sopwith Snipe that, on October 27, 1918, he fought his most famous air battle, witnessed by thousands of Allied infantry below. The lone Canadian took on fifteen Fokkers of Jagdgeschwader 3, and in the ensuing dogfight, wounded and barely conscious, he crashed into no man's land below. When Barker woke up ten days later in hospital in Rouen, he was told that he had shot down four Fokkers with two as probables — and he had been awarded the Victoria Cross. In 1924, William Barker would help found the

Major W.G. Barker in a Sopwith Camel aircraft of No. 28 Squadron, Italy, 1918.

RCAF and later become vice-president of Fairchild Canada. It was in this job, while test-flying a Fairchild from Rockliffe, Ottawa, that he would meet his death on March 12, 1930.

VICKERS VIMY

On Blackmarsh Road in St. John's, Newfoundland, is a monument to where Canada (although Newfoundland would not join for another thirty years) first reached out over the Atlantic to Europe. The aircraft used wasn't Canadian, and neither were its pilots. But fifteen years after man had first flown, Captain John Alcock and Lieutenant Arthur Whitten Brown shrank the Atlantic Ocean between Canada and the mother country, cementing old ties and fostering British-Canadian unity.

The end of the First World War meant a surplus of pilots and aircraft, both available to commercial operations at a fraction of their worth. If some of the unemployed pilots took to stunt flying and barnstorming, others challenged the distances between the Old World and the New, between Britain and the colonies. Before the war, Lord Northcliffe, the London press baron, had offered a prize of £10,000 for the most direct flight across the Atlantic Ocean. In April 1919, the prize money was raised to £13,000, and the wartime development of aviation made such a flight feasible — almost.

Four British teams competed for the Northcliffe prize, each funded by an aircraft manufacturer: Sopwith, Martinsyde, Handley Page, and Vickers. At its narrowest point, the Atlantic Ocean is two thousand miles of unpredictability, and with the prevailing winds blowing from west to east, Newfoundland was the most suitable jumping-off point. The islanders must have wondered what hit them. From never having seen an aircraft to becoming hosts to four of the latest! Three of the teams set up operations at the Cochrane Hotel, St. John's (Vickers was a late entry). Although not competing, the U.S. Navy was also nearby, at Trepassey Bay. That spring, it was conducting a naval

Assembly of Vickers Vimy aircraft of Captain John Alcock and Lieutenant Arthur Whitten Brown at Lester's Field, St. John's, Newfoundland, June 1919.

Courtesy of the National Archives of Canada

exercise, flying three Curtiss NC-4 flying boats across the Atlantic, guided by no less than sixty destroyers strung out at fifty-mile intervals all the way. The NC-4s had to fly from one ship to another, and on May 16, the naval flying boats took off, making for the Azores. As well, there was also a U.S. Navy dirigible in Newfoundland ready to make the crossing. Not for another three decades, when Gander would become the crossroads of transatlantic travel, would Newfoundland attract so much attention in the aviation world.

Media attention on the British teams was high, not in the least because Northcliffe's newspaper made sure it was. After the grief of the war years, when the casualties at Ypres, Jutland, and the Somme had filled the headlines, the public was in a euphoric mood, eager to see what aviation could do. With the promotion that winning the race would bring, each manufacturer coerced its team to get into the air before the others did. Perhaps unprepared, the Martinsyde and Sopwith entries took off, to defeat, on May 18. Caught in a crosswind, the Martinsyde crash-landed on the airfield itself. The Sopwith fared better, but in the middle of the ocean its engine overheated and the crew ditched near a small ship, which rescued them. Only the giant Handley Page bomber was now left, and it looked as if it would carry the day.

Then on May 26, the Vickers Vimy arrived in crates. Christened "Atlantic," it was assembled at Pleasantville, the former Sopwith airfield, with some of the crew sleeping in the crates at night to guard it. News came on May 31 that the surviving U.S. Navy NC-4 had arrived at Plymouth, completing the Atlantic crossing by air. The Vickers crew were more concerned about the size of the Pleasantville field, and on June 4, the Vimy was flown to Lester's Field, at the west end of St. John's.

Courtesy of the National Archives of Canada

The Vickers crew was ex-RFC pilot Captain John Alcock and navigator Lieutenant Arthur Whitten Brown. Both had dreamt of accomplishing this while prisoners of war. By coincidence, when a flight instructor, Alcock had taught the young Canadian aviator J. Erroll Boyd to fly. Their

Vickers Vimy aircraft of Captain John Alcock and Lieutenant A.W. Brown ready for transatlantic flight at Lester's Field, St. John's, Newfoundland, June 14, 1919.

29

aircraft, the Vimy, had been designed by Vickers as a strategic bomber and first flew from Weybridge on November 30, 1917, too late to be used operationally in the war. For its day it was a large aircraft, with a 69-foot wingspan and 43-foot fuselage. Each of the successive marks used different engines, the Vimy IV settling on twin 12-cylinder, 350-horsepower Rolls Royce Eagles, each driving four-bladed wooden propellers measuring 10 feet in diameter. The advantage of the Eagle was that it could fly for one hundred hours without servicing— the average time then was thirty hours. To make it across, extra fuel tanks were provided, for a total of 856 gallons of fuel. There was a simple radio, which Alcock and Brown used, but it died three hours into the flight. Creature comforts for the men were side-by-side seating on a hard bench, sandwiches made by a Miss Agnes Dooley, coffee, bottles of beer, and whisky. For mascots, each brought along a toy cat.

Daily Vickers and Northcliffe harassed them with cables demanding to know what the delay was. They replied that the weather was wretched, heavy rain and ice that could quickly force an aircraft down. Finally on June 14, in a lull between storms, after a favourable weather report that was sketchy at best, Alcock and Brown pronounced themselves ready. Locals pushed the overburdened Vimy to the far end of the hastily prepared runway. The Eagles roared into life and the plane trundled along, building up speed, Alcock just clearing the stone wall and some trees at the runway's end. It was 2:00 P.M. local time and 1612 GMT.

As they passed over St. John's harbour, ships' sirens blew the aircraft a farewell and Alcock noted that the altimeter read 1,083 feet. The first four hours were calm, the altimeter was a steady 4,000 feet, and the sky was clear. Then a fog bank enveloped the Vimy and, blinded, they flew on, with Alcock taking the Vimy higher so that Brown could get a fix by the sun. Suddenly, sounding like machine gun fire, the right engine burst into flame just as the electrical heating in their suits gave up. The aircraft turned into a spiral towards the ocean. By the time they came out of the fog they were just above the waves, only sixty-five feet below. The engine's exhaust had burnt off but the Eagle was behaving, so Alcock opened the throttle, swung the aircraft back on course, and took it up to 7,200 feet. The crisis behind them, the sandwiches were eaten and the beer drunk.

Vickers Vimy aircraft of Captain John Alcock and Lieutenant A.W. Brown taking off on transatlantic flight, Lester's Field, St. John's, Newfoundland, June 14, 1919.

The long flight continued into the night, with Brown staying warm by pumping gas from the main tank in the fuselage into the secondary ones by the engines. Now they saw the stars and, using the sextant, plotted their course, calculating where they were — still only halfway across. Then, with the first signs of dawn, they were thrown into a storm, causing the Vimy's nose to dive and the ocean come up to them again — this time vertically. Alcock straightened the Vimy in time to fly into a snowstorm. Snow clogged up the air intakes and air filters on the Eagles' carburetors and the engines began to falter. Blinded by it, Brown climbed out onto the nose, made his way to the wing, and, with a knife, chopped the ice off the inlets, then went back over the nose to do the other wing. While he did this, Alcock tried to keep the aircraft balanced, knowing that one hasty move would cause Brown to slip to his death, and without him, his own demise would surely follow. Brown performed this manoeuvre four times until the snowstorm ended.

Dawn appeared and the sun caused the snow around the two men to melt, leaving a wet cockpit. Numb, wet, and deafened, they saw land at 0825 GMT. It definitely wasn't Galway, their intended destination. But Brown recognized the top of Connemara and they descended to the little town of Clifden. Looking for a suitable landing, Alcock spotted the radio station and circled it. Just below it was a green meadow, and, ignoring the waving men at the radio station, he made for it. This was Derrygimla Moor, a local swamp, and the men had been trying to warn the Vimy off landing there. At 0840 GMT on June 15, 1919, the Vimy touched down, causing two deep tracks in the green meadow before burrowing its nose into the bog. Just as the noise of the Eagles died, a Mr. Taylor ran up to the fliers and asked where they had come from. Strangely enough, America was the reply. After a flight of sixteen hours and twenty-seven minutes, the Atlantic had been conquered.

Alcock and Brown were knighted and the Vimy was repaired and put on display at the Science Museum in South Kensington, London, where it is today. Vickers did capitalize on the Atlantic's fame, and the RAF ordered many more, using them in the Middle East as troop carriers. On December 18, 1919, Alcock was killed in an air crash in Normandy. He would have liked to see his former student, J. Erroll Boyd, on October 10, 1930, become the first Canadian to fly the Atlantic. By then, the ocean had been defeated by a German zeppelin and by Charles Lindbergh. But Alcock and Brown were the first. They had opened up the world for us.

Curtiss JN-4 aircraft C332 of the Royal Flying Corps of Canada, Camp Borden, Ontario, c. 1917.

CURTISS JN-4 CANUCK

Forever associated with barnstormers, wing-walkers, and Mack Sennett silent movies, the Curtiss JN series not only trained British, Canadian, and American airmen during the First World War and thrilled thousands of fairgoers after it, but it was also responsible for beginning the aviation industry in Canada.

This American icon was actually designed by a Briton, B. Douglas Thomas, whose Model J aircraft was powered by the Curtiss 90-horsepower water-cooled OX-5 engine. Thomas is also credited with introducing the British tractor concept to Glen Hammond Curtiss, who had until then had built pushers. The Curtiss engineers added to his J a control column that allowed the forward and aft motion to operate the elevators, and a little wheel on it that steered the rudder. The company already had a Model N, and Curtiss merged both designs together for the JN series, the initials lending themselves to the name Jenny.

Curtiss was a Western Union messenger boy who realized that if his bicycle could be motorized it would save time and energy. The lightweight combustion engines that he then built at Hammondsport, New York, made him first one of the earliest motorcycle racers in the United States and then an aeronautical engineer. Invited by Alexander Graham Bell to join the Aerial Experiment Association in 1906, Curtiss's engines powered the aircraft at Baddeck, Nova Scotia, as well as the first aircraft to fly in other parts of Canada (the 1910 Minoru Park racetrack in Vancouver and the Winnipeg Industrial Exhibition). At this early stage in aircraft development, there was a definite divergence in where the powerplant should be. Following the Wrights, North American aeronautical engineers favoured the pusher concept; emulating Bleriot, the Europeans liked the tractor. In 1907, Curtiss left the AEA and began his own aircraft manufacturing company at Hammondsport; he was later joined by J.A.D. McCurdy, who flew his pusher planes on several record flights. By 1913, Curtiss was building flying boats or hydro-aeroplanes, and it was then that Thomas introduced the tractor J model.

Nothing speeds aircraft development quicker than a war, and when the First World War began, the Curtiss Aeroplane & Motor Company, now in Buffalo, New York, looked to exploit the Allied need for pilots. The Admiralty especially wanted as many airmen as it could get and liked the idea of setting up flying schools in Canada, equipped with Curtiss JN trainers. But the United States was neutral, and worse, Orville Wright was threatening legal action against any manufacturer who made any sort of wing control device, claiming that the Wrights had invented it. Canada was excluded from this ban as it had no aviation industry, and to circumvent the Wright claim Curtiss hired his old AEA friend J.A.D. McCurdy to begin an aircraft-manufacturing subsidiary in Toronto and train pilots at Long Branch in the city's suburbs. The Curtiss/McCurdy plan fitted together neatly. The Curtiss plant at 20 Strachan Avenue (near the Canadian National Exhibition grounds) made the ailerons for JN-2s and 3s that were brought across from Buffalo, the former for the Spanish air force and the latter for the Long Point flying school. On May 10, 1915, the Curtiss Aviation

School began operations on Toronto Island using Curtiss flying boats, also soon to be built at Strachan Avenue, and on July 14, 1915, Anthony H. Jannus flew the first JN-3 at Long Point. But the slaughter on the Western Front demanded more pilots than the Curtiss school could produce — hundreds more, and quickly. On December 15, 1916, the British government took over all pilot training and, borrowing $1 million from Ottawa, purchased the Curtiss operations and renamed the Strachan plant Canadian Aeroplanes Ltd. It opened flying training schools for the RFC across Canada, the first at Camp Borden in Ontario, all using Curtiss aircraft.

Hilliard Bell stands in front of a Curtiss JN-4 prior to a June 1917 training flight.

Courtesy of the DND

The historic hangars of Camp Borden, Ontario, the birthplace of the RCAF.

By January 1917, the prototype Curtiss JN-4 was completed and was tested at Long Branch by Bertrand B. Acosta. Completely built in Toronto, only its Curtiss engine had been imported from the United States. Called the Canuck, it was distinguished from the American-built Jennies by the ailerons on the lower wings. Eventually, Canadian Aeroplanes would build 1,210 Canucks, with spares for 1,700 more. Although there were other manufacturers (like Ericson Aircraft, which built thirty-seven between 1920 and 1926, and JV. Elliot, which assembled seven from 1925 to 1927) aircraft manufacturing in Canada really owes its beginning to the aileron workshop at Strachan Avenue.

Even before the Armistice, JN-4 trainers were being used in less than orthodox ways. In the summer of 1917, Lieutenant Ervin E. Ballough of the RFC stunted in a JN-4 over Deseronto, Ontario, by leaving his cockpit. That winter, the first ski flying experiments were carried out by 44 Wing RFC with JN-4s. All airmail in Canada dates from June 24, 1918, when Captain Brian Peck and Corporal E.W. Mathers flew from Montreal to Leaside, Toronto, in a JN-4 with a small sack of mail. On May 5, 1919, Ballough carried 150 pounds of raw furs in his JN-4 from Toronto to Elizabeth, New Jersey, inaugurating cross-border commercial flights.

But the lure to stunt in a JN-4 was irresistible. By 1919 there were hundreds available — some at fifty dollars each — so that anyone could become a barnstormer, and did. Travelling air circuses made up of unemployed ex-RFC pilots in Canucks sprouted across North America, and no county fair or city exhibition was complete without an aeronautic display. The public soon tired of air races and wanted stunts: loops, stalls, wing-walking, jumping between Jennies, and parachuting. F.H. Ellis, Canada's first aviation historian, made the first parachute jump from an aircraft in Canada when he leapt from a JN-4 over Crystal Beach, Ontario, on July 5, 1919. Few barnstormers actually wore parachutes— all stunts had to be done close to the ground where the audience could see them, and it was a credit to the aircraft's inherent stability and strength that it allowed this abuse to take place. The JN-4's projecting king posts were perfect for wing-walkers to do handstands and even pretend to play tennis on the top wing.

But the JN-4 Canucks were more than circus sideshows, performing historic feats as well. In what is the first recorded instance that an aircraft was used in police work, W.R. "Wop" May flew a detective from Edmonton to Coalbranch, Alberta, in August 1919 to catch a criminal. On August 7, 1919, the first flight across (or through) the Rockies took place when Ernest C. Hoy flew a JN-4 from Vancouver to Calgary via Vernon, Grand Forks, Cranbrook, and Lethbridge. The first airmail from Truro, Nova Scotia, to Charlottetown, Prince Edward Island, on September 29, 1919, was by a JN-4. Finally, the JN-4 was rightly honoured when one belonging to the Aerial Service Company of Regina, Saskatchewan, became the first civil aircraft to be registered in Canada as G-CAAA.

In the years that followed, there were to be many other aircraft built in Canada, but the JN-4 Canuck holds the honour of being the first.

VICKERS VEDETTE

Decades before the Norseman and the de Havilland Beaver, there was the Vickers Vedette. With its polished mahogany plywood body, it looked more like a rich man's wooden speed boat than a bush plane, but the little Vedette was the first aircraft designed and built in Canada for local conditions.

As late as 1925, Canada was an unknown country. Its coastline had been mapped by and for mariners, and apart from the railway line, much of the interior was still unexplored. After the war to end all wars, the government of the day saw little need for maintaining an air force — unless it could be gainfully employed to map and photograph the landmass. Other sectors were also crying out for civil use of military aircraft: forestry protection, geological surveying, aerial spraying, medical rescue, and transport of Indian treaty agents. These were all seen as tasks that the RCAF could perform — if it had suitable aircraft.

The Canadian Air Force and later the RCAF was equipped with the aircraft it had received as part of the Imperial Gift, a far from ideal situation. The Curtiss flying boats built to hunt German submarines and the DH4 fighter aircraft designed to meet the Red Baron were unsuited to the peacetime conditions of endless forestry patrols, transport, and fire suppression. What was needed was a locally built machine configured for Canadian conditions. As there was no national industry to provide for such an aircraft in 1924, the RCAF, familiar with Vickers machines, approached the British company to ask if they would be interested in building a flying boat in Canada. Vickers wanted a steep $15,000 for each aircraft, but they offered to send chief designer Wilfrid Thomas Reid to set up a facility in Montreal, and they guaranteed the first flying boat ready within six months. Reid obviously had the plans for a

Vickers Vedette.

flying boat at hand, for in July 1924, a model called the Vedette was being put through wind tunnel tests at the University of Toronto, the first such tests to be conducted in Canada. The contract was signed in August, and Reid set up his workshop at the Vickers shipbuilding yard in the port of Montreal. By November 4, the prototype Vedette G-CYFS was test-flown by Flight Officer W.N. Plenderleith of the RAF, and between 1924 and 1930, forty-four Vedettes in five marks were built for the RCAF. The plane carried three: the pilot and engineer, with the observer/photographer up front in a round, bathtub-type seat. It wasn't an ideal bush aircraft as it was fragile and could not be used in the winter. But the Vedettes were lighter and more manoeuvrable than the Curtiss HS2Ls. They also proved to be reliable, easy to fly, and, it was said, able to take off from a lake in under ten seconds.

Vickers also made sixteen Vedettes for commercial operators. The first, G-CAFF, built for Fairchild Aerial Surveys Ltd. on May 10, 1926, was also the first commercial aircraft to be built in Canada. Different engines were tried out: the Rolls Royce Falcon, the Wolseley Viper, the Wright J-4, and the Lynx IV. The main difference between the civil and military Vedettes was that the RCAF standardized its engines with the British-built Lynx and the civil versions used the American Wright Whirlwind J-4. In 1925, former First World War air ace Don McLaren opened his own air company, Pacific Airways, at the edge of Bute Street in False Creek, Vancouver. The company lived on fisheries patrols and aerial spraying using a collection of flying boats, one of which was the Vickers Vedette Mark V G-CASW. McLaren's crewman Gordon Ballentine recalled that they had been spoiled by their Boeing B1E flying boat, which was enclosed and had upholstered seats. With the Vedette, he said, "You sat out where God meant airplane drivers to sit; out in the noise and wind." The Vedette had a horrible motion in rough air. In the summer of 1928, on a flight from Prince Rupert to Queen Charlotte City, Neville Cumming, Gordon Ballentine, and Alf Walker lost their way in cloud between two islands and their Vedette hit the trees on Porcher Island. The radial engine Wright J-5 fell forward into cockpit, causing some injury to the pilot. Eventually all three were rescued, but the Vedette was a write-off.

The Vedette has the distinction of being the aircraft that the first Canadian to save his life by parachute would jump out of. On May 17, 1929, Canadian Vickers test pilot Colin Jack Caldwell bailed out of a spinning Vedette over Montreal. Caldwell was living on borrowed time, for on June 20 of that year, he was killed when his Fokker hit high-tension wires over the St. Lawrence.

A Vedette is prepared for launching. Designed by W.T. Reid, the Vedette was a commercial success.

Later, when the RCAF relinquished forest fire patrols, provincial governments would purchase their Vedettes to conduct their own. Reid left Vickers in 1928 and began his own aircraft company at Cartierville, building a light aircraft called the Reid Rambler. The Depression killed off his business as it did Vickers' aviation venture until 1936, when Ottawa gave it the contract to build the Stranraer flying boat.

The little flying boats remained in use somewhere in the Canadian bush until 1941,

Vickers Vedette.

and six were even exported to Chile. Of the several Vedettes converted to amphibious status, one was the former RCAF aircraft CF-AIS. It was bought for a nominal sum in 1934 by the government of Manitoba, registered as CF-MAG, and used on forest protection. When in 1937 its Wright J-6-9 engine failed on a flight, it sank into a swamp and disappeared for years. Recovered by the Western Canada Aviation Museum in Winnipeg, the Vedette was rebuilt and is today the only one of its kind in existence.

DE HAVILLAND 60 MOTH

This is the connection between a man and a place. Although he never saw Borden, Ontario, Geoffrey de Havilland would have a profound effect on it, as he would on such airfields worldwide. His DH 60 Moth would be used at Camp Borden to train hundreds of Canadians in both the military and commercial fields.

Born in 1882, the son of the Reverend Charles de Havilland, Geoffrey was fascinated by anything mechanical. While still at the rectory, he built a petrol engine to power his bicycle. His first job was at the Wolsey Tool and Motor Car Company, and when his grandfather left him £1000, the young man was able to finance his dream — designing and building a flying machine with the help of his brother-in-law, Frank Hearle. Several trials followed, and on September 10, 1910, de Havilland aircraft was first successfully flown. During the First World War, Geoffrey worked first at the Royal Aircraft Factory and then the Aircraft Manufacturing Company at Hendon.

His first aircraft, the DH 1 (he numbered each of his aircraft), was almost a copy of an existing Fe2. It was his second design that "made" him. The DH 2 was a pusher — the engine was in the rear, which allowed the pilot to fire his machine gun unobstructed, an advantage that the RFC's first ace, Lanoe George Hawker, used to good effect. Unfortunately the pusher concept made it slow and an easy target — Hawker fall victim to the guns of Manfred von Richthofen, the Red Baron. The DH 2s were soon relegated to training schools, but Geoffery de Havilland had gotten his chance. Post-war, in a small field at Stag Lane, Edgeware, he began his own company, which, whether through Hawker Siddeley, British Aerospace, or de Havilland Canada, would supply the world with innovative aircraft.

DH 60 Moth.

Camp Borden rightly calls itself the birthplace of the RCAF, having begun as a First World War RFC training school; it was here in 1927 that Wallace Turnbull perfected his invention, the variable pitch propeller. Four years earlier, Borden had been selected as the only training institution in Canada for young men who wished to be pilots in the Canadian Air Force. Chosen from universities, they came to Borden to drill and study the theory of flight, map reading, aerial photography, basic engineering, and signalling. For this they were paid the handsome sum of $3.50 a day and provided with uniforms, quarters, rations, and medical and dental treatment. If they passed the ground school, they sat in the back of a First World War Avro 504K, trying to comprehend what their instructors did with it. Eventually, the students graduated and were assigned to some forlorn outpost in the bush or to Vancouver, where they were taught seaplane flying. Canada needed pilots for civil activities such as fire and fishery patrols, and the Borden graduates were not going to fly fighters or bombers but become uniformed bush pilots.

When international events forced the RCAF to assume a more military stance, the government wisely opened the Borden program to commercial pilots, teaching them blind flying and radio. Many of those who took the course, like Robert Randall, George Reid Phillips, and Wilfrid Wop May, went on to greater things, eventually becoming members of Canada's Aviation Hall of Fame.

By 1928, the Avro 504s were falling apart, and when in March of that year the de Havilland workshop at Weston, Ontario, assembled its first DH 60 Moth, the RCAF saw a replacement. On August 3, 1928, the first DH 60X Moth was delivered to Camp Borden. Geoffrey de Havilland had developed the 60, or Gipsy Moth (the first of the immensely successful Moth family) in 1925. Powered by a 60-horsepower in-line Cirrus engine (later changed to a 100-horsepower Gipsy engine in 1928) the X version would have a split undercarriage as opposed to the straight axle type. The aircraft was easy to fly and affordable and became a favourite of wealthy private owners such as the Prince of Wales and long distance fliers Amy Johnson and Francis Chichester.

In Canada, as part of a government-assisted scheme, DH 60s were used to set up flying clubs — the Montreal Flying Club's first aircraft was the DH 60 C-GAKD. On

Courtesy of the Schade family

Between the wars, the RCAF trained many bush pilots at Camp Borden, Ontario, teaching them "blind flying" and radio. Many of those who took the course, like Robert Randall, George Reid Phillips, and Wilfrid "Wop" May, eventually became members of Canada's Aviation Hall of Fame.

January 28, 1929, the first aircraft of the new CF series was the Hamilton Aero Club's DH 60 CF-CAK. The Department of Fisheries sent one on the Hudson Straits Expedition, and the Ontario Provincial Air Service bought four. One of the many historic flights that the DH 60 was used on culminated on November 18, 1930, when a Newfoundland Airways aircraft flew from Toronto to St. John's, Newfoundland, bringing the first air-mail from Canada to the British colony.

DH 60 Moths.

At Borden, the DH 60s were soon the mainstay of the Advanced Flying Training Course, and in 1929 they were first ski-equipped, allowing for more realistic winter training. The last Avro 504 was retired from Borden in March 1930, and the following year another of Geoffrey de Havilland's creations appeared when thirteen Puss Moths were bought for the RCAF. Intended for advanced instruction and instrument training, they and the DH 60s would be utilized for the blind flying course at Borden, a first attempt at teaching Canadian pilots, both military and civil, instrument flying.

So successful were the DH 60s that in 1938, de Havilland took the giant step of building aircraft at Downsview, Ontario, beginning with its successor, the DH 82A or immortal Tiger Moth.

There are still DH 60s flying today — built in France and Australia — and Borden is still fulfilling its role as a Canadian Forces military base, both proving that a good idea is timeless.

FORD TRIMOTOR

Lumbering, slow, and noisy. A flight in a Tin Goose, said one journalist of the day, was like going a few rounds with a heavyweight boxer, unarmed. For days after such a flight, passengers remained deaf from the engine noise and ill from the smell of vomit and engine exhaust poisoning. But whatever its faults, the Ford Trimotor was the first true American airliner, the forerunner of all the DC-3s, 707s, and 747s. Three such Fords were destined to play their part in the birth of commercial aviation in Canada.

In 1924, when William Stout built the 2-AT, an all-metal monoplane, he hoped that the American government would purchase it to fly the mail. It didn't, but Stout got the attention of someone equally as powerful — Henry Ford. The auto-maker was looking to start an air express company to ship parts around his various factories, and on July 31, 1925, he bought Stout's company, retaining him to redesign the aircraft. The autocratic Ford then took over and gave the aircraft three 200-horsepower Wright air-cooled radial engines. When the ungainly aircraft, now called the 3-AT, barely lifted off the ground and was mysteriously destroyed in a fire, Ford dismissed Stout and set his own company's engineers onto the project. They had never designed an aircraft before but were more afraid of their boss than such a handicap. Fortunately for them, a convenient example literally came out of the sky. The polar explorer Admiral Richard Byrd was a friend of Henry Ford and brought over the Fokker V11 trimotor that he was taking on his next trip to the Antarctic. Legend has it that while the two great men had lunch, Ford's engineers crawled all over the Fokker, measuring and photographing it in detail. Using this they came up with the template for the Tin Goose, the 4-AT, which first flew on June 11, 1926. All subsequent versions, the complete 5-AT to 10-AT series, were reconfigured from this basic design.

Ford Trimotor.

It was the height of the Roaring Twenties and venture capital was plentiful. If everyone, it seemed, had a Ford Model T in their garage, every air company wanted their own Ford Trimotor on the tarmac. Canadian-born Clement Keys ran Transcontinental Air Transport and offered, in combination with the railroads, flights between New York and Los Angeles in four days. Passengers would fly in the Fords by day and sleep in the Pullman railcars at night. At a time when flying was thought so dangerous that no insurance company had a policy covering it, Keys reassured his customers by hiring as advisors the two most famous pilots of the day — Charles Lindbergh and Amelia Earhart. Lindbergh, impressed with the all-metal, three-engine German Junkers G 31 monoplane, recommended the Ford 5-AT, and Keys placed an order large enough to guarantee the 5-AT's success, with Lucky Lindy flying his personal Ford while developing the transcontinental air route.

If passengers were tempted onboard TAT's Fords with such luxuries as windows that opened, toilets (a hole in the rear), and in-flight meals cooked by Fred Harvey, they soon discovered that the noise from the Pratt & Whitney engines made conversation impossible and the sliding windows were not for ventilation (as the aircraft were always freezing) but to throw the "barf bags" out of. What frightened them even more was the Ford's unusual rudder and elevator control system: for easy maintenance all cables ran from the pilot's yoke to the tail on the outside of the fuselage, and in flight these thumped alarmingly against the tin sides. The Depression killed off the rail-air service (the TAT initials soon stood for Take A Train), and in 1933, after a total of 199 Trimotors were built, Henry Ford turned his back on aviation.

The first Ford Trimotor (indeed, the first multi-engine aircraft) to come to Canada was a 4-AT bought by British Columbia Airways that arrived in Vancouver on August 7, 1928. Flown by Ted Cressy and Hal Walker all the way from Detroit and registered as G-CATX, it was to be used on the Vancouver-Victoria-Seattle service, scheduled to begin August 16. On August 25, with five passengers and a fox terrier on board, Hal Walker and Hal Carson took off from Lansdowne field in Victoria and flew into a fog bank near Port Townsend, Washington. A fisherman saw the aircraft turn violently, a wing tip hit the water, and the Ford cartwheel into it. There were no survivors, and British Columbia Airways closed down.

The second Ford trimotor in Canada belonged to the American-owned Colonial Air Transport and landed at St. Hubert Airport, Montreal on October 1, 1928, to inaugurate the Canada-U.S. airmail service. While no commercial air company in Canada needed so large an aircraft, the penurious RCAF did scrape up enough to buy a used Ford 4-AT in June 1929 for crop and forest dusting. At a time when the federal government was considering a non-military role for the RCAF, combating wheat rust on the Prairies or Hemlock Looper disease in Quebec by aerial spraying was seen as the most cost-effective way of recouping the taxpayers' investment. In any case, its three engines and two-thousand-pound payload made the Ford safer and more economical than the little Keystone Puffer biplane then being used by the RCAF. Registered as G-CYWZ, the Ford also took on transport duties, becoming the support aircraft for the Siskin aerobatics team on cross-Canada tours and taking government VIPs to see the R100 in August 1930. When the RCAF put it up for sale in 1933, the only Canadian who could afford the asking price of $55,000 was Harry Oakes, the mining millionaire, who registered it as G-CARC and kept it at Fort Erie airport near his Niagara Falls mansion.

Enter smooth-talking aviation entrepreneur Grant McConachie. Appealing to Oakes's memories of how he made his fortune prospecting in the bush, he managed to talk him into giving up the Ford for $2,500. McConachie employed the aircraft's hauling capacity to good advantage for his company, United Air Transport, then based in Edmonton. Typically, on the flight home from Fort Erie to Edmonton, he piled twelve friends into it and almost crashed the overloaded aircraft on the way home. The Ford was used to fly sacks of frozen fish (up to 3,600 pounds a flight) from Peter Pond Lake, Saskatchewan, to the railhead at Cheecham. But being McConachie, fish hauling was not adventurous enough and what followed was pure grandstanding — McConachie style.

Ford Trimotor aircraft 7683 taking part in the inauguration of the Canada–United States airmail service, St. Hubert, Quebec, October 1, 1928.

On May 16, 1935, milking as much publicity as he could from the trip, the aircraft still reeking of fish, he took off from Calgary Airport at 9:30 A.M., aiming for Vancouver. Carrying a single passenger, he intimated to the media that this was the start of commercial air flights over the Rockies. But the Ford wasn't built to cope with ice over the mountains and its broad, corrugated wings soon collected enough to make the controls first sluggish and then inoperable. McConachie barely made it down to a small airstrip at Grand Forks, British Columbia, near the American border. There the whole town came out to welcome the huge aircraft — still the largest in Canada — and after refueling, he continued on. No one had thought to notify the press in Vancouver of the delay, and the assumption that the Ford was now

Ford Trimotor.

Ford Trimotor.

Courtesy of the Schade family

a twisted wreckage on a peak in somewhere in the Rockies was broadcast on the radio, adding to the McConachie story. Someone heard the Ford fly over and phoned the news in so that when he landed at Sea Island Airport, there was a huge crowd waiting. It was not the first nor the last time Grant McConachie would make the headlines.

Sadly, commercial flights across the Rockies would have to wait another three years, and McConachie returned to Alberta for more fish hauling, with some barnstorming in the summer. He did use the plane (now registered as CF-BEP) to pioneer the first airmail service between Edmonton and Whitehorse, Yukon, on July 5, 1937. But while the Ford could take pontoons, it could not be adapted for skis and was of little use in the winter, away from airfields. With the advent of aircraft like the Norseman, which accommodated wheels, skis, and floats, McConachie hoped to get rid of the Ford — but for a decent price. Once more the gods smiled on him. It was June 1939, and the RCAF had moved its most modern fighter aircraft, Hawker Hurricanes, to Sea Island, Vancouver, as escorts during the Royal Visit. They shared the airport with the ancient Ford, now part of McConachie's Yukon Southern Air Transport as it flew freight to Fort Nelson through the summer. On June 24, the tired old Tin Goose sat placidly outside the Vancouver Airport administration building waiting for the next load when a Hurricane pilot, attempting to take off, froze at the controls and slammed into it. While the young man survived, the old aircraft did not. McConachie sued the Crown for loss of freight contracts and a valuable aircraft. His luck held as the judge ruled in his favour, and he was awarded $52,000. He also got $5,200 from the insurance company for the Ford, twice what he had paid for it. In its death, the ancient airliner/aerial sprayer/fish hauler had given Grant McConachie a head start on the road to one day becoming president of Canadian Pacific Airlines.

As to the other characters in the aircraft's history: TAT combined with Western Airlines to become Trans World Airlines. The embittered Harry Oakes moved to the Bahamas, where in 1943 he was murdered by a still unknown assailant. Ted Cressy went on to fly for several other airlines, and the aero caterer Fred Harvey, realizing that the public would eat anything, opened a chain of hamburger outlets. Strangest of all, Bill Stout, the original all-metal aircraft designer, reappeared in 1955 to build copies of the Ford Trimotor. When Henry Ford's heirs threatened to sue, Stout renamed his aircraft the Bushmaster. In 2002, there were eight Ford Trimotors still flying somewhere in the United States, with eleven in storage or museums.

H.M. Airship R100 approaching mooring mast upon completion of transatlantic flight. St. Hubert, Quebec, August 1, 1930.

THE R100

To greet thee, Canada most fair;
A glorious land that is to be,
A nation rich, and strong, and free;
A Silver Star, upon an azure sea.[1]

It didn't have wings and it didn't go further west than Niagara Falls. It wasn't even the first dirigible to cross the Atlantic — the German Graf Zeppelin had done so twice before. But the British dirigible R100's visit to Canada in the summer of 1929 affected the nation as much as Charles Lindbergh's flight had the United States two years before. It wasn't just the R100's triumphant ocean crossing — this was taken to be the awakening of the Dominion of Canada. As Fred Rowlett's contemporary poem hoped, the R100's arrival foretold a new era. Whatever happened after, the country would never be the same.

Using the new design methods of Barnes Wallis and Neville Shute-Norway, two ships, the R101 and the R100, were built by Britain in 1927, one by the government and the other by Vickers. They were very different, and they would travel in opposite directions on their maiden international voyages: the R101 was to travel east to India, and the R100 west across the Atlantic. However, both used hydrogen, recognized for its flammability. Helium was safer, but the United States had a monopoly.

---

1. *R100 in Canada.* Erin: Boston Mills Press, 1982.

If any single Canadian was responsible for the visit of the R100 it was Prime Minister William Lyon Mackenzie King. He agreed to participate in the Imperial Airship Scheme, a British globe-girdling plan to use long-range dirigibles to bind the Empire closer to London and away from American influence. In April 1927, a British Air Ministry team toured Canada for the location of an airship mast and terminus. They drove around Toronto, Halifax, and Ottawa (where the Connaught Rifles Ranges and Rockliffe airfield were considered) before deciding on the village of St. Hubert on the south shore of

Dining room photo taken by Canadian passenger on R100 tour.

Montreal. When inevitably the press asked of Lindbergh's recent flight, the British airship officials passed it off as a wonderful stunt and no more, prophesying that transoceanic travel belonged to airships alone. After all, what transatlantic aircraft could carry one hundred passengers who enjoyed the luxuries of a dining room, showers, and bunk beds in flight? (The only pleasure that airship passengers couldn't indulge in was smoking.)

As Canada's contribution to the Airship Scheme, in 1928 a mooring mast that cost $376,000 was erected at St. Hubert and completed on March 31, 1930, the airport leveled and paved as the first federal airport in Canada. In return, King wanted a demonstration of the airship's prowess, ordering that to get as much publicity as possible, it circuit Ottawa, Quebec City, Toronto, Hamilton, and Niagara Falls. Other Canadian cities begged to be included; Halifax offered financial incentives, while the city fathers of Vancouver were told that it was impossible because of the Rockies.

For weeks before the newspapers and movies played up the R100's interiors. Its cubic capacity of more than 5,000,000 cubic feet, its double staircase leading down to the dining room, the panoramic windows, and the two-tier promenade deck became familiar to most Canadians. The meals compared with those of a luxury liner, yet so smooth was the airship's motion that no one was ever seasick on board. The effects of the Wall Street crash the previous October hardly dampened the public's enthusiasm; if anything, the R100's visit distracted from the grim realities that the country was beginning to undergo. The intended visit from its base at Cardington, north of London, to St. Hubert was planned for the summer of 1930 and created great excitement among Canadians (far surpassing

TAMING THE SKIES
A Celebration of Canadian Flight

the general election), especially in the sales of radio sets. Radio stations from coast to coast and below the border clamoured to broadcast the every move of the dirigible; the *Toronto Daily Star* newspaper even sent its star sports announcer, a young Foster Hewitt, to Montreal to do a play by play.

King would have liked the R100 to tour while he was campaigning, but the weather for a crossing wasn't favourable until July 29. When it slipped from its mast at Cardington at 2:48 GMT that day with thirty-seven officers and men and seven passengers (one of whom was author Neville Shute-Norway), Canada was just digesting the results of the election. R.B. Bennett's Conservatives had defeated MacKenzie King's Liberals, and the Imperial Airship Scheme was being denounced as King's parting excess, one that he learned from in years to come.

Forty-six hours and twenty-seven minutes later, the R100 reached Newfoundland, and at dawn on July 31 it began radioing Montreal. At St. Hubert it was estimated that the R100 would now arrive at 4:00 P.M. that day, and police sirens combined with military bugles and train whistles to awaken the citizens of Montreal as police and troops at the airport prepared for it. It was the middle of the summer holidays and it seemed that every schoolboy who owned a bicycle made it to St. Hubert that morning. At noon the official welcome party arrived from Ottawa by the RCAF Ford Trimotor. The main hangar at St. Hubert was draped with Union Jacks and echoed with the clamour of two hundred reporters either playing cards or typing out their copy, which would be sent around the world by forty telephone and wireless operators who had been specially brought in. All over Montreal, families sought the best picnic sites from which to watch for the dirigible's arrival.

Then at 2:00 P.M. Foster Hewitt broadcast that the R100 had reduced speed over the mouth of the Saguenay River after a squall had torn open the lower fin. As the fifteen-foot gash was being repaired with fabric and dope (a glue), the estimated time of arrival was pushed back to 8:00 P.M. That evening, as disappointed Montrealers returned home in

Courtesy of the National Archives of Canada

R100 illuminated while at St. Hubert Airport.

the darkness, the R100 encountered a thunderstorm and reduced speed further as it pushed its way through. But at about 2:35 A.M., August 1, all over the city people were awakened by the throb of its engines and, rushing out into the streets, saw in the moonless night the R100's navigation lights and windows. The dirigible circled Montreal, awaiting the light of dawn to dock. As the sun rose, it approached the mooring mast from the east at a height of five hundred feet, the British ground crew, which had left on May 16, making the mooring cable secure. The six Rolls Royce Condor engines stopped at 5:23 A.M. and the cable was made taut. In the fading night, movie cameras and searchlights played on the silver envelope, the black letters G-FAAV and R-100 clearly seen. The total crossing of 3,364 nautical miles had taken 78 hours and 49 minutes, including the time spent on repairs. Compared with Lindbergh's Ryan monoplane the R100 was slow. He had flown from New York to Paris in thirty-three hours — but then he had flown alone, without a dining room, bunk beds, and baths.

Later generations of Montrealers who would celebrate the opening of Expo 67 in their city would be told by their grandparents that it was nothing compared with the welcome they had given the R100.

Promenade room on the R100.

Greetings and best wishes were everywhere — on billboards, stamps, signs, illuminated newspaper advertisements, postcards, lapel buttons ... there was even a hit tune written especially for the occasion, "Here's to the R100 ...we are with you as you soar," which was broadcast from the radio station at the Windsor Hotel daily.

Beginning at 10:00 A.M. that day, trains from the CN station on Guy Street shuttled the public to St. Hubert's main entrance every fifteen minutes. Over 70,000 people streamed through the entrances that first day, and in the twelve days that followed 530,000 Canadians and Americans joined them. It was a masterpiece in crowd control, thanks to the Army and Mounties — who strictly enforced the no-smoking regulations. The airship's interior was admired by eight thousand people, and it was reported that a twenty-four-year-old millionaire movie producer named Howard Hughes phoned from New York, wanting the R100's crew to come to Manhattan, where his aviation movie *Hell's Angels* was about to open. The crew themselves were guests of honour at the huge banquet that Montreal's flamboyant mayor Camillien Houde threw for eight hundred guests on August 8 at the Windsor Hotel, the dessert bombe glacée R100 created especially for the occasion.

On Sunday, August 10, Houde arrived at St. Hubert with his motorcycle police escort to see the R100 set off on its Canadian tour at 6:18 P.M. It passed over Lac des Deux Montagnes, Oka, and then Plantagenet while its passengers, the Ottawa VIP party, dined and played bridge. In Ottawa that evening, the open spaces around the Parliament buildings and Majors Hill filled with spectators who were entertained with the music from the Parliament building carillon, which had been another of Mackenzie King's projects. There was a bright moon, and at 9:50 P.M. the lights of the R100 were spotted, causing every automobile horn and railroad whistle to drown out the carillon's rendition of "Rule Britannia" and "O Canada." From a window in the Château Laurier the newly elected Prime Minister R.B. Bennett (who lived at the hotel) and Ottawa mayor Frank Plant watched and greeted the R100 via radio. The airship circled Ottawa and the suburbs, passing over Rideau Hall, Hull, the Connaught Ranges, Carleton Place, and Smiths Falls (where many climbed the city water tower for a better view) before making for Toronto. The only place it did not grace was Kingsmere, Mackenzie King's estate, where that night the former Prime Minister sat, embittered. He had asked DND that his wish be relayed to the R100 but (he suspected) it must have been conveniently lost on the way. He had brought the dirigible over, lost the election, perhaps, because of the expense — and now Bennett was taking the glory....

Its engines roused Oshawa at 4:10 A.M. and Toronto Harbour at 4:45 A.M., where the ship's whistles made sure that Torontonians knew the R100 had arrived. Then she made for Niagara Falls, New York, returning to Canadian territory to circle St. Catharine's and Hamilton. By the time she had returned to Toronto, its citizens had woken up fully and flooded the building roofs downtown. As she passed over City Hall at 9:28 A.M. the breakfasting passengers could not help noticing the huge Union Jack and fifteen-foot sign "Welcome R100" on the driveway below. That day, August 11, a restaurant called the R100 opened on Yonge Street and Briar Hill Avenue, keeping the name until 1937. The headline in the *Toronto Daily Star* said it all: "Industry At Standstill as R100 Visit City."

The R100 left St. Hubert at 9:30 A.M. on August 16, arriving at Cardington 57 hours and 56 minutes later, a flight made speedier because of the Gulf Stream. It was generally thought that this was to be the start of a scheduled service between Britain and Canada. The crew was given little rest, as they were also responsible for the sister ship R101, which was being prepared to leave for India. With its destruction on October 5, 1930, over Beauvais, France, all future flights were cancelled and the R100 was deflated on December 11, 1930. As many of those killed at Beauvais had been part of the R100 crew, the Canadian government ordered that flags be flown at half mast throughout the country and memorial services were held in Montreal and on the steps of the Toronto City Hall where the "Welcome R100" sign had been.

The Depression killed off any subsequent airship schemes, and when he was re-elected Mackenzie King had the mooring mast at St. Hubert dismantled in 1938. Never to be used again, it had now become a hazard to the Trans Canada Airlines Lockheed Electras then operating from the airport. With the discovery of helium deposits in Canada, it was hoped that the Dominion would buy the R100, but that was not to be and it was broken up and sold for scrap in February 1932. Gradually, in the bleak years of Depression and war that followed, the memory of that sunny August when the R100 visited and all things seemed possible for Canada faded away.

Northrop Delta.

NORTHROP DELTA

Aircraft designer Jack Northrop pioneered the use of all-metal stressed skin fabrication and the multicellular wing. The late 1920s and early 1930s were the era of speed, when the fastest aircraft competed — in Europe for the Schneider Trophy and in the United States for the Thompson and Bendix trophies. Northrop's series of powerful, fast aircraft, the Alpha, Beta, Gamma, and Delta, were sought after by racing aviators like Howard Hughes and Roscoe Turner and explorers like Admiral Richard E. Byrd.

The third in the Northrop series, the Gamma was the first to come to Canada. The low-wing, cantilever monoplane with its 14-cylinder, 785-horsepower Wright Whirlwind engine and undercarriage in streamlined trousers had been designed as a long-range, high-ceiling, high-speed transport — and it looked the part. It was built to order in 1932 for aviator Frank M. Hawks, who used it to break distance and speed records on behalf of his sponsor, Texaco Skychief. On August 4, 1933, he flew it from Regina, Saskatchewan, to Bridgeport, Connecticut, a distance of 1,700 miles, in 7 hours and 50 minutes. Possibly training for the Bendix Trophy race, on August 25 he flew from Vancouver to Quebec City via Kingston and Montreal in 17 hours and 10 minutes. Hawks received maximum publicity for this in Canada, and in 1934, looking for a photo-reconnaissance aircraft, the RCAF made overtures toward Northrop.

But the Gamma was to be a one off, and the only one built exploded in mid-air in 1936. By then Jack Northrop's eyes were already on the regional airline market, for which he built the Gamma 2B, a transport version that Lincoln Ellsworth and Canadian pilot Herbert Hollick-Kenyon flew across the Antarctic. Northrop progressed from the only 2B (which Ellsworth left in Australia for the Royal Australian Air Force) to the Delta, stretching the fuselage so that it could accommodate two pilots and eight passengers. Unfortunately, even as he was designing it, the American government forbade the airlines to use single-engine aircraft.

Now only the RCAF was interested in Northrop's aircraft, and it wanted the Gamma, not the Delta. But as it wasn't being made anymore, the Canadian government had to settle for the Delta, and in 1936 Canadian Vickers was given the licence to build four aircraft. The redesigned racing aircraft was to become the RCAF's maid of all work. The floor was strengthened to take Fairchild cameras and a door was installed on the side for freight. Two were even armed with a machine gun to be fired from a dorsal rear open

Northrop Delta.

Courtesy of the DND

hatch and given racks to carry 250-pound bombs under the wing. Provision was made for wheels, skis, or floats, and between 1937 and 1940, seventeen Delta II's were built for use as bombers, reconnaissance, transport, and, what must have been suicidal in a single-engine aircraft maritime, patrol. The Department of Transport even suggested that Trans Canada Airlines equip itself with Deltas to fly over the Rockies

With the war, as inadequate as the Deltas were, they served for a short while as Canada's only long-range maritime patrol aircraft. The aircraft would account for the first RCAF casualties of the Second World War when on September 14, 1939, one flown by Wing Officer J.E. Doan and Corporal D.A. Rennie disappeared on a flight over New Brunswick, the remains not found until July 1958. The Delta's curtain call came in 1941, when 8 (BR) Squadron operating Deltas from Sydney, Nova Scotia, on the pretence of submarine patrol gratefully accepted the first Bristol Bolingbrokes as replacements.

SUPERMARINE STRANRAER

At a time when the PBY Catalina and Sunderland were operating, the Stranraer was a quaint, Jurassic-looking aircraft. Yet it had been designed by the same British company famous for those Schneider Trophy winners and the rakish Spitfire. It had so many bracing wires that it was referred to as The Whistling Birdcage. The Stranny looked as if it was left over from the First World War. In fact, its parents were the Imperial Airways Kent flying boats and the Shorts Singapore series of the mid 1930s — what the media of the day called Ships with Wings. It may have looked frail, but its hull was encased in Alclad sheeting, and sandwiched between the fabric-covered dura-lumin wings were two 920-horsepower Bristol Pegasus X radial engines that gave it the lift and reliability needed to serve Canada through the war — and after.

The twin-engine biplane flying boat had originally been named the Southampton Mark V by Supermarine, which built twenty-three for the RAF. The name was changed to Stranraer in 1935, and the following year the Canadian government contracted Vickers to build forty for the RCAF at its De Maisonneuve Street plant in Montreal. The Stranraers would be constructed in the same riverside shops where the Vedettes had been made a decade earlier and the PBYs would be five years later. With double-decker wings and a nautical bow, it was reminiscent of a ship under full sail, but it was Canada's only maritime patrol aircraft and when King George VI and Queen Elizabeth toured Canada in the early summer of 1939, their liner was grandly escorted out of Halifax by the Stranraers of No. 5 (GR) Squadron.

With a crew of six and three Lewis guns — in the bow, dorsal, and tail — it had a maximum bomb load of 1,000 pounds. It had a maximum speed of 165 miles per hour and an effective range of about 720 miles and was never

Supermarine Stranraer of 5 (BR) Squadron over sailing vessel, April 3, 1941.

designed to take on German raiders or for long-range convoy escort. But in the very first RCAF action of the war, No. 5 Squadron Stranraers from Dartmouth, Nova Scotia, flew patrols on September 10, 1939. Six days later, when the first convoy of ships for Britain put out of Halifax, it was preceded by No. 5 Squadron Stranraers. The flying boats searched for German submarines off the harbour before the convoy's departure and then provided an escort for 250 miles. A pattern was established in the initial months of the war when the flying boats would take off from Dartmouth at first light to patrol over outgoing convoys, land at Sable Island in the afternoon to refuel, and then rejoin the convoy, making it back to Dartmouth before nightfall. After that the convoy was bereft of air cover until it came within range of the RAF in Northern Ireland. The aircraft had an endurance of about nine hours, which allowed it patrols of five hours and thirty minutes duration, giving it three and a half hours to find a location and then fly back to base.

A No. 5 Squadron Stranraer featured in a little-known sideshow far from Halifax harbour. Flight Lieutenant Len Birchall was flying coastal patrols out of Dartmouth, and in May 1940, when a declaration of war with Italy seemed imminent, he was tasked with locating all Italian merchant ships still in the St. Lawrence. On June 1, Birchall discovered the freighter *Capo Lena* off Anticosti Island, making a dash for the open sea before the declaration could be announced over the radio. He shadowed the ship all day waiting for the hostilities to begin but at dusk was forced to return to base. The next day he returned and continued to follow it, but once more there was no declaration and he returned home, allowing the *Capo Lena* into the Atlantic. Finally, on June 10, Canada declared war on Italy and Birchall once more scoured the St. Lawrence for Italian ships. This time he was in luck.

Courtesy of Canadian Airlines.

Jim Spilbury's Queen Charlotte Airlines Stranraer.

The *Capo Nola* had just left Quebec City and was making a run for the Gulf. Its crew must have watched incredulously as Birchall's Stranraer bore down on them like a dowager aunt pretending to be a police constable, evoking as much menace as it could. The Italian captain was sufficiently frightened by the pilot's intent to run his vessel aground on a sandbank. Birchall set the Stranraer down near it until a Royal Canadian naval vessel showed up. The crew of the *Capo Nola* became the first Italian prisoners of war, and Birchall was soon to be promoted and posted to 413, an RCAF Catalina squadron in the Shetlands.

As the Douglas Digbys and Lockheed Hudsons became available, the Stranraers were moved to other locations: Shelbourne, Nova Scotia, and Jericho Beach, British Columbia. After the war, the RCAF wanted to dispose of its twenty-four Stranraers lying at Jericho Beach. Ashton James "Jim" Spilsbury, owner of a Cessna Crane that serviced the Queen Charlotte Islands, raised enough money to buy two of the Stranraers for $25,000 each. Spilsbury converted them to twenty-passenger aircraft and launched his Queen Charlotte Airlines with services from Vancouver to Skidgate and Sandspit. The first QCA Stranraer, chris-

Supermarine Stranraer flying boats of 5 (GR) Squadron, RCAF, escorting S.S. **Empress of Britain** *carrying King George VI and Queen Elizabeth back to England. Dartmouth/Halifax, Nova Scotia, June 15, 1939.*

tened the Haida Queen, took to the air on March 2, 1946, for Prince Rupert. The Stranraers and later Cansos of Spilsbury's QCA (sometimes called Queer Collection of Aircraft) provided a cheap, reliable air service on the British Columbia coast, putting the local shipping line out of business. When Spilsbury was bought by Pacific Western Airways in 1955, QCA was the third largest airline in Canada.

There is a single Stranraer in existence today. To the shame of Canadians, the remaining former RCAF and QCA Stranraer was snapped up by the British and is on display at the RAF Museum at Hendon.

AIRSPEED AS-10 OXFORD

Known to all who flew her as the Ox-box, the Oxford was developed by the British aircraft manufacturer Airspeed from its light airliner, the Envoy. This twin-engine monoplane aircraft had been built in 1934 as a feeder or regional airliner to Imperial Airways, and as A.V. Roe did with its Anson when war looked likely, Airspeed offered it to the military as an aircrew trainer.

The Oxford was first supplied to the Central Flying School in November 1937 as a bombing, navigational, and gunnery trainer, and by the time the war began, there were four hundred in service. The plywood-covered aircraft with its two Armstrong Siddley Cheetah X engines was docile enough, and the Mark I had an Armstrong Whitworth dorsal gun turret fitted to it for gunnery practice. It became the standard aircrew trainer for the RAF, and when the British Commonwealth Air Training Plan was signed, it was the principal British aircraft contribution to it. The contract to build more Oxfords was shared between Airspeed, de Havilland, and Standard Motors, and the aircraft was shipped out to BCATP flying schools in Canada, Australia, New Zealand, and Southern Rhodesia.

Besides training, Oxfords also served as air ambulances and transport. The first shipment of twenty-five Oxford Mark Is came to Canada in 1939, to be followed by 606 Mark IIs in 1941 and 188 Mark IVs the following year. Only the Mark Is had gun turrets, and by 1943, the Armstrong Siddley engines had been replaced with two 455 Pratt & Whitney Wasp Juniors. By the end of the war the Oxfords had become the airfield "dogsbodies."

Airspeed Oxford Mark I fitted with an Armstrong Siddeley Cheetah X powerplant during April 1942. The RCAF had twenty-seven aircraft of the mark (registration numbers 1501-1025, A89, and A101) on strength from May 8, 1939, to April 26, 1944.

SHORTS SUNDERLAND

When it appeared in 1937, the Air Ministry Public Relations called it Queen of the Seas and The Flying Battleship. To the Luftwaffe, because it was portly and bristled with guns, it was the *Fliegende Stachelschwein* (Flying Porcupine). To everyone who was fortunate enough to fly in one or even see one taxiing, the Sunderland represented the very acme of seaplane development, combining grace and power with majesty.

In the late 1930s, as tensions grew in Europe, the RAF realized that it would soon need a long-range convoy escort and maritime reconnaissance aircraft, not only in the home waters but in the Mediterranean and Far East as well. The winning bid by Shorts to the Air Ministry specification for such a long-range flying boat, the first Sunderland flew on October 16, 1937. The company had used experience gained from its Imperial Airways Empire boats to build a streamlined, high-wing monoplane that, except for a change to Pratt & Whitney engines, required little modification over twenty-one years of service. Between 1937 and 1946, 749 Sunderlands were built, mainly at the Shorts factory in Rochester and later at Windermere, Dumbarton, and at Harland and Wolff in Belfast.

The only anomaly in the classic design was that the engines seemed out of line. The C class flying boat wing put the aircraft's centre of gravity too far aft, and rather than redesign the wing, the Shorts designers swept it back, giving the engines a curious splayed look canting outwards and not in the direction of the aircraft. It cruised at a regal 110 knots thanks to the built-in tailwinds caused by the engine nacelles. For its day, it was a massive craft — four Bristol Pegasus Mark I nose and tail turrets and soon a mid-upper as well, stations for eleven crew, and a complete galley. If watching a Sunderland take off as it cut through the waves was an unforgettable experience, being the pilot of one was equally mem-

Shorts Sunderland.

Courtesy of the Public Archives of Canada

orable. To take off, the craft had to be kept in a planing condition until the required speed was reached. For this the outer engines were kept at full throttle. At the same time the Sunderland had to be held straight, which in a rough sea was dodgy; if the nose fell, the aircraft ploughed into a wave, if it rose, the plane might be thrown prematurely into the air. The pilot's manual instructed: "When the nose rises and spray is clear of the inner propellers, open the inner engines to full throttle. At 50,000 pounds take off speed is 75 knots." No crewmember ever forgot the resounding bumps every times the aircraft was thrown into the air and came down as it built up enough speed and got into planing attitude.

Courtesy of the National Archives of Canada

Flight deck of a Sunderland, No. 422 Squadron. Flight Lieutenant McBrien (navigator).

Because Sunderland pilots combined seamanship with airmanship the atmosphere on board was closer to that of a Royal Naval vessel than an RAF aircraft. As flights were usually of a twenty-hour duration the crew had to be self-sufficient in body and soul, and so far from base the captain or skipper was as autonomous as any on the bridge of a ship. The galley in the lower hull forward of the bomb compartment was well-equipped: it had Primus stoves, a Clyde Cooker (oven), kettle, pans for stew and frying, teapot, cutlery, and drying cloths. The Catalina and Liberator crews who flew patrols of longer duration without such amenities looked on their Sunderland colleagues with envy. This really was an Empire flying boat going to war.

When war did come, Sunderlands were well armed, equipped with gun turrets, depth charges, bombs, and mines. The depth charges were carried inside the fuselage and winched out under the wing on patrol. The flying boats were operated by the RAF, the RCAF, the RAAF, the RNZAF, the Free French, and the Royal Norwegian Air Force. Two RCAF squadrons flew Sunderland Mark IIs and IIIs — 422 squadron was based at Pembroke Dock and 423 squadron at Castle Archdale, Lough Erne, Northern Ireland. It was while flying a Mark III G-George from Lough Erne on May 12, 1943, that Flight Lieutenant John Musgrave sighted U-456. As it dove, the Sunderland dropped two depth charges, scoring a hit, so that the U-boat could be finished off by escort vessels. In battles with the U-boats, 423 lost six aircraft and forty-nine aircrew and 422 lost eleven aircraft and fifty-two aircrew. It was a heavy price to pay for the seven recorded incidents of RCAF Sunderlands sinking German submarines.

George O'Neill, who flew as a rear gunner in a Sunderland, remembers:

When I joined a Sunderland crew in the spring of 1944, straight from Air Gunners school, I was impressed with its sheer size which, with an upper deck, enabled crews to move around freely, and swap duties thus keeping fresh and vigilant during flights of up to sixteen hours. My first Sunderland flight was from Lough Erne in Northern Ireland. It was exhilarating to see the spray thrown up as we gathered speed then got up on the step — the forward fuselage — and thrust itself into the air. We flew down a neutral corridor over Eire to Donegal Bay and thence into the Atlantic. Convoy protection and anti-submarine patrols were long and tiring, but we usually had good rations — eggs and steak if we were lucky! And hot sweet tea in giant flasks to fortify us. There were two Primus stoves in the Galley but when the hatches were opened to throw out smoke or flame floats — to check wind speed and direction — the stoves would be blown along the bench and the cooking pans scattered!

Operationally, the Sunderland had a crew of eleven — three pilots, a navigator, three wireless operators, two flight engineers, and two air gunners. The wireless operators and engineers also had gunnery training. The armament consisted of fourteen machine guns — eight in the three power turrets, four forward firing guns in the hull — operated by the pilot; these guns were all Browning .303 and in the galley hatches were mounted two Vickers .3 guns. My position was in the rear turret cramped between four guns. A wonderful vantage point to spot shipping and aircraft. After four or five hours I rotated with the mid-upper gunner or took a short break in the galley.

With such heavy armament, the Sunderland could take care of itself. In May 1943, an Australian-crewed Sunderland fought off eight Junkers JU 88s in the Bay of Biscay. At this time there were twenty operational Sunderland squadrons flying in every theatre of war.

During the winter of 1944 we were based in Scotland; the weather was generally bad. The aircraft trots were about a mile from the service jetty, and on a rough day we arrived at the aircraft cold and wet. The bilges then had to be pumped dry, and the moorings made ready to slip. Once airborne many of the guns had to be mounted and loaded — not easy in flight.

Often on returning at last light from a long flight, Flying Control would say that a high wind was expected, and a gale crew was to be left aboard. This consisted of a pilot, engineer, wireless operator, and gunner. The engines were run to keep the aircraft up to the buoy. Alternatively, a skeleton crew was left to help refuel the aircraft. Neither duty to be relished!

In Scotland we converted to the Mark V Sunderland, which had many improvements, including the more powerful Pratt and Whitney engines and two heavy additional machineguns in the waist. We flew our new Mark V out to the main Flying Boat base in the Far East, at Koggala in the south of Ceylon — now Sri Lanka. There we flew from a lake which was not very long and if not airborne soon the aircraft might hit the trees at the end of the lake! On days when there was a flat calm, the marine craft would criss-cross the lake to make waves to help us get airborne.

The war in South East Asia continued for about three months after V-E day, and anticipating peace after the Atomic Bomb was dropped, our aircraft were modified to carry ex-POWs from Singapore. All the armour and armament were stripped out, and fitments were installed in the waist to take stretchers. All Sunderlands in the Far East located at various bases round the Indian Ocean were to be involved in creating an air bridge for ex-POWs. Our Flying Boat Tender, the SS *Manela*, sailed to Seletar, Singapore, late in August to organize the operation and provide Flying Control. All aviation fuel was brought in 50-gallon drums in tank landing craft and a labour gang of one hundred coolies was recruited to offload the fuel and transfer it to the marine craft refuellers by bowsers. A marine craft section refurbished these craft — left behind over three years previously.

The first flight of six Sunderland aircraft from 205 and 230 squadrons flew to Seletar early in September, and on their return reported organizational problems there. As I was the only aircrew officer not required for flying, being an air gunner, the Station Commander at Koggala sent me to Seletar to sort out the problems and control the various activities there. On my way to Singapore we overflew the Allied invasion fleet — Operation Zipper — a most impressive sight. We landed on September 15th, the day that Mountbatten took the Japanese surrender in Singapore. The main problems were soon sorted out and the air bridge came fully into operation. About a fortnight later, RAF Dakotas were able to operate from the adjoining airfield,

Courtesy of the National Archives of Canada

Sunderland W6027, No. 422 Squadron, RCAF, in flight showing radar aerials.

and thus many ex-POWs were airlifted from there to Ceylon and India. Whilst I was busy dealing with day-to-day problems I managed to fit in two flights. Our aircraft was packed with stretcher cases and those who could walk. As many of the ex-POWs were suffering from several diseases and most had dysentery these twelve-hour flights were difficult! The air and ground crews were first class and we were all very pleased and proud to take part in this huge errand of mercy. I was twenty years old at that time.

Six Mark III Sunderlands were converted by Shorts during the war for Transport Command to be used as passenger flying boats for a service from Poole to Lagos. Immediately after the war six more were converted for BOAC's network. Called Sandringhams, with nose and tail turrets removed, they were refitted with two decks, the upper for a dining room and bar and the lower for seats and berths to carry up to twenty-four day or sixteen night passengers. When this ended in 1949, the Sandringhams were purchased by other carriers, as were Sunderlands by other air forces, in particular the South African and French.

With the exception of one in the United States, the few remaining Sunderlands are no longer in flyable condition. It is a pity that future generations will not know the power and grace of one creaming through the waves on takeoff. It truly was the last instance when Britannia ruled the waves.

HANDLEY PAGE HAMPDEN

The Flying Suitcase, the Hambone, and the Panhandle were some of the more polite names that its crews gave it. If its poor armament and drafty, freezing cabin were not bad enough, the maximum width of its fuselage was a claustrophobic three feet, preventing all movement during flight. An injured crewmember was almost impossible to reach; worse, if he happened to be the pilot, it was too late. After enduring heavy losses early in the war, the RAF Bomber Command passed its Hampdens over to Coastal Command, the Australians, the RCAF, and the Red Air Force.

When first operated in 1936, the Handley Page Hampden was a revolutionary concept. The pod and boom design was an innovation in Britain at that time, as were the Hampden's wings, with their automatic leading edge slots that allowed for low landing speeds. Designed by Handley Page's Dr. D.V. Lachmann, it was a fast, compact fighter/bomber, which was why it had a forward-firing Browning gun. Twin boomed and pencil thin like the Dornier Do. 17 (for which it was often fatally mistaken during the war) it was faster and more manoeuvrable than the Vickers Wellington. It could carry twice the bomb load of a Blenheim twice as far.

The body was so narrow that the fuselage was built in halves, the fittings installed into each half — the only way to get them in — and then joined together. It goes without saying that adding more equipment, like gun turrets or armour, or even doing repairs internally was almost impossible. There was also little room to replace the hand-held, drum-fed Vickers with heavier machine guns. The biggest irritant must have been instructing on the Hampden. There was no side-by-side seating, and trainee pilots learned to fly by standing behind the instructor and watching

Handley Page Hampden.

him as he did circuits and landings. When it was the trainee's turn the situation was reversed and the instructor breathed down the poor novice's neck — woe betide him if he brought the Hampden in for a hard landing and the instructor hit his head on the ceiling!

Meant to be mass-produced and easily built, the Hampden was made by a number of companies. Besides the 550 that Handley Page turned out at Cricklewood, 770 were built by English Electric at Preston, 150 at Shorts in Belfast, and 160 at Canadian Associated Aircraft Ltd. at Malton and St. Hubert. The RAF entered the war with 212 Hampdens in 8 squadrons and sent them out on daylight raids over Germany before realizing their inadequacies. The 1,000-horsepower Bristol Pegasus engines were not powerful enough to escape the fighters, and the guns in the nose and rear dorsal and ventral positions had little traverse. If the defensive firepower of the hand-held Vickers guns was negligible, the fixed machine gun in the nose was useless — the pilot had to turn the whole aircraft to aim it. The few RAF Hampdens that returned from the daylight raids were put on minelaying duties and bombing the Channel ports at night.

It was in one of these raids, on September 15, 1940, that a Canadian pilot flying for the RAF was awarded the Distinguished Flying Cross. Twenty-one-year-old Pilot Officer Clare Connor from Toronto, with 83 Squadron based at RAF Scampton, was bombing the Antwerp docks. The Air Ministry Bulletin 1839 takes up the story:

> His first run over the target was inaccurate and no bombs were dropped. On the second approach, at 2,000 feet, the aircraft was subjected to intense fire from the ground, but the attack was pressed home successfully. During this attack, the bomb compartment in the aircraft was shattered and a fire started which quickly spread to the wireless operator's and rear-gunner's cockpit. The port mid-wing and the tail boom were damaged. Shellfire pierced the port rear petrol tank causing grave risk of the fire spreading, and the starboard tank was also pierced. The navigator and rear gunner abandoned the aircraft, but the wireless operator/air gunner remained and succeeded in controlling and eventually extinguished the flames. In spite of the condition of his aircraft and knowing that he had neither a navigator, rear gunner, or normal wireless facilities, Pilot Officer Connor succeeded in flying back to his base and landing without further damage. He displayed the most outstanding coolness, courage, and devotion to duty.

Connor was awarded the Distinguished Flying Cross, and his wireless operator, eighteen-year-old John Hannah, the Victoria Cross. Recommending the Canadian for the DFC, Air Vice Marshall A.T. Harris wrote, "The condition of this aircraft has to be seen to be believed." Hannah never flew again, but Connor did and was killed on November 3, 1940, when his Hampden crashed into the sea. By September 1942, all Hampdens were phased out of Bomber Command and given completely over to Coastal Command to be converted into torpedo bombers.

In Canada, the manufacturing of Hampdens was part of an industry program to one day make Shorts Stirling heavy bombers. Of the 160 Hampdens built, 84 were sent to Britain by sea and the remainder to Victoria Airport (Patricia Bay) B.C., to be flown by RAF 32 OTU squadron. Perhaps because they had some rudimentary heating installed, the Canadian-made Hampdens became part of the Allied military aid to the Soviet Union. On September 2, 1942, flown by aircrew from RAF 144 squadron and the RAAF 455 squadron, the Hampdens were

Hampden bomber built at Canadian Associated Aircraft, St. Hubert, Quebec, August 8, 1940.

dispatched to Soviet bases near Murmansk. Some disappeared along the way, possibly because their crews had become disoriented flying so close to the magnetic north. The Canadian Hampdens were handed over to the Red Air Force and, perhaps to their relief, the RAF and RAAF personnel were sent home by sea. The Russians used them as torpedo bombers to protect the convoys, keeping a special lookout for the German battleship *Tirpitz*.

Luckless RCAF squadrons were also relegated Hampdens for use as torpedo bombers, making anti-shipping strikes with them deep into the Bay of Biscay. To avoid the Ju-88s that patrolled the area, the Canadians would fly at less than one hundred feet; remaining at that altitude (without an autopilot) for as long as eight hours in the cramped fuselage was a nerve-wracking, fatiguing exercise. The RCAF found the Hampdens were particularly vulnerable during torpedo strikes as they were required to fly very low and to very slowly launch a torpedo to have any hope of success. Yet somehow, on August 2, 1943, an RCAF Hampden torpedoed the German submarine U-706. When the more powerful Bristol Beaufighters were available, the squadrons gratefully traded in the few Hampdens that were left for them.

As is the case with unpopular aircraft, few Hampdens survived the breaker's yard during the war. Fortunately for future generations one did — in Canada. On November 15, 1942, while on torpedo practice near Patricia Bay, British Columbia, a Canadian Hampden crashed into the water. It was salvaged by the Canadian Museum of Flight in Langley, British Columbia, where it rests today, *sans* engines and partially restored.

NOORDUYN NORSEMAN

The Swedish Air Force museum at Malmen and the Western Canada Development Museum at Moose Jaw share a treasure. In both are Norsemen — sturdy, yellow, snub-nosed air ambulances. The only difference is that the Swedish one has red crosses on its wings and fuselage. Because of the polio victims it flew to hospitals and the midwives it took to isolated prairie communities, the Canadian Norseman CF-SAM is part of Saskatchewan folklore, and many adults today call themselves CF-SAM babies. One would like to think that both examples are what Robert Noorduyn had in mind when he designed the aircraft.

When Noorduyn set about designing a bushplane, he asked the men who flew them what they wanted to see in one. Aircraft until then had been built with the American airline market in mind, not the Canadian North. When working for fellow Dutchman Anthony Fokker, Noorduyn had met pilots from Canadian aviation companies who came to the Fokker plant in Teterboro, New Jersey, to buy aircraft like the Fokker Universal and then adapt them for conditions in the far North. What was needed was a bushplane, a flying truck that was rugged, economical, and dependable.

In 1934, Noorduyn moved to Montreal and formed his own aircraft company in the suburb of Cartierville, where, on the old polo grounds, various small manufacturers had set up. Work was begun on the prototype Norseman, a high-winged monoplane with a wide and strong undercarriage that was interchangeable for wheels, skis, and floats. The engine was the 420-horsepower Wright Whirlwind. Registered as CF-AYO, it was put on floats and made its first flight from Pointe aux Trembles, Montreal Island on November 14, 1935. Although it was the prototype and still undergoing testing, Dominion Skyways bought it in January 1936. The second aircraft, CF-AZA, flew on May 2,

Rugged beauty: the classic Norseman.

Courtesy of the Noorduyn family

1936, it too finding a ready buyer. The bush operators liked the ten-foot cabin and the fact that in later models the payload was three thousand pounds. The only complaint was that the Norseman was underpowered, and Noorduyn replaced the Whirlwind with the 450-horsepower Pratt & Whitney Wasp, and in later versions the 600-horsepower P & W Wasp. Three Norseman Mark IIs and one Mark III were built, but it was the IV that he turned out in large numbers (ninety-three), and the first customers were bush air companies like Canadian Airways. In 1938 the Royal Canadian Mounted Police purchased one, and the RCAF, gearing up to expand, bought eight to be used as trainers. Once war was declared the Air Force purchased a further forty-seven Norsemen for military liaison duties.

With the world's attention focused on the outbreak of the Second World War, the Norseman might have ended its days in obscurity, as a good Canadian-built aircraft, had not the United States begun preparations to enter the war in 1941. President Roosevelt had promised to supply the British with aircraft — bombers and fighters — and the urgency to get them to Britain, compounded by the initial successes of German submarines in the Atlantic, gave birth to ferrying them over. Staging airbases in Labrador, Newfoundland, Greenland, and Scotland were set up to pass the aircraft through — almost from factory to front line. The isolation of these areas called for aircraft that could perform transport and SAR in climatic conditions that were much like the Canadian North. The Norwegian pilot Bernt Balchen, who worked for Fokker and came to Canada in 1927 to fly Universals in the historic airlift to Fort Churchill, recommended the Norseman to the United States Air Force. The Americans bought six to maintain the ferry route through Greenland, calling them YC-64s. Once the United States was in the war, the orders multiplied. Norsemen were ideal for whatever the Army Corps of Engineers had in mind: the construction of the Alaska Highway, island bases in the Pacific, duties in North Africa, and work in Europe, where after D Day they served just behind the front lines as flying ambulances. Under Lend-Lease, in 1943, the Royal Australian Air Force received fourteen Norseman from the U.S. Military Defense Aid Program. By 1945, the American military had purchased 762 Norsemen, now in its Mark VI version.

Courtesy of the National Archives of Canada

Noorduyn Norseman aircraft in production for the Royal Canadian Air Force. Montreal, Quebec, March 1941.

While many bush airlines used the Norseman, none equaled the number that Canadian Pacific Air Lines had at various times. To keep the airline's vital bush network active, in 1944 the government allocated to it eight Norseman VIs (CF-CPL to CF-CPS), and once the war had ended it bought seven (CF-CRC to CF-CRF, CF-CRS to CF-CRU) from the Crown Assets Disposal. When DC-3s became available and the airline expanded to urban areas, Canadian Pacific sold off their Norsemen to bush operators like Mont Laurier Aviation, Central Northern Airways, and Ontario Central Airlines.

Post-war, the glut of military surplus Norsemen on the market depressed prices, and in 1946 Canadian Car & Foundry bought Noorduyn out. In May 1953, a consortium headed by Robert Noorduyn repurchased the company and continued production, making the last of 904 Norsemen in 1959.

B-17 FLYING FORTRESS

When Catherine Wyler convinced Warner Brothers to remake her father's wartime film *Memphis Belle* in 1989, the villagers at Binbrook, Lincolnshire, where the movie was shot, were treated to sights and sounds that they never thought possible. As they had forty-four years before, United States Army Air Corps B17s were once more lumbering over the British countryside and sometimes coming into land as the script dictated, with smoking engines and feathered props. If the Spitfire epitomized Britain in the Second World War, the B-17 Flying Fortress epitomized the United States. Here was an all-American aircraft, and even today, to see a Fortress in a museum is to hear Glenn Miller's "Moonlight Serenade" and imagine its eight-man crew on board — gum-chewing Iowa farm boys or Hollywood stars like Clark Gable and Jimmy Stewart, both of whom served in Forts.

When Boeing entered the multi-bomber field in 1935, the only experience it could draw on was its Boeing 247 airliner. Using it as a model, it began work on a four-engine Model 299 on August 16,1934. The Army Air Corps at first thought it impractical and far too complex to operate, and when the prototype crashed on October 30, 1935, it seemed to bear them out. But they reckoned without the public. The size and armament of the bomber, crowned by the inspired name Flying Fortress, got it media attention far out of proportion. Here was American military might on patrol, its rugged construction and heavy firepower ensuring that it lived up to its name. If early on in the war its deficiencies and operational misuse made it less of a fortress, these were corrected by 1944, the year of its greatest use. Even the manufacturing pool that was assigned to build B-17s (Boeing, Vega, and Douglas), better known as BVD (a brand of underwear), would give it a Norman Rockwell folklore appeal. In total 12,726 B-17s of all models were

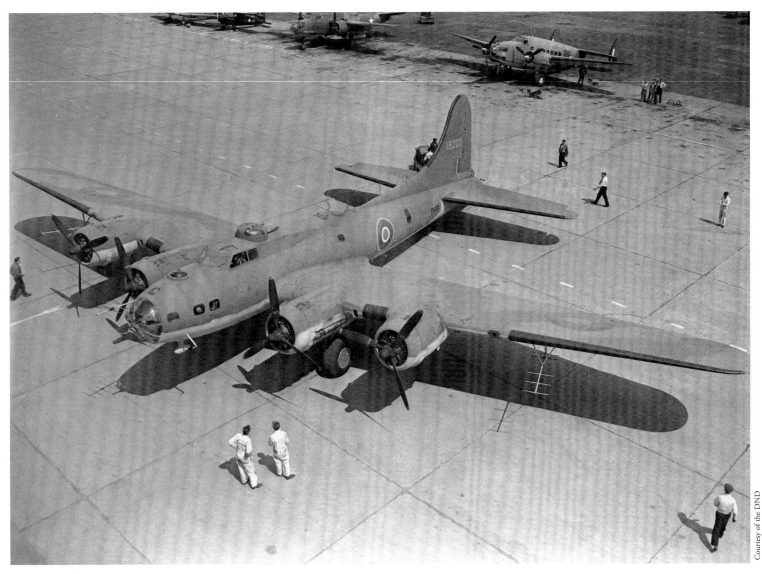

B-17 Flying Fortress operated by RCAF.

Courtesy of the DND

built, and they served in the USAAC, the RAF, and the RCAF. The high density of their use meant several ended in up in enemy hands (the Luftwaffe alone had forty B-17s that it used for espionage work) or in neutral countries (seven became Swedish airliners).

The B-17 came to the RCAF in 1943 not as a strategic bomber but as a transatlantic mail aircraft. Operation Mailcan was to carry mail to Canadian troops in Europe, and the only long-range, four-engine transports that were available were six used B-17s. From December 6, 1943, in RCAF Serial Nos. 9202-9207, three B-17Fs and three B-17Es were taken on strength by 168 Heavy Transport Squadron at Rockliffe. The first B-17 flight took off on December 17, piloted by Wing Commander Robert Bruce Middleton. Middleton had extensive experience in such flights, having flown for Imperial Airways in 1936 and been a founding member of Trans Canada Airlines in 1937, captaining the first TCA flight from Vancouver to Winnipeg the following year. In October 1943, Middleton was put in command of 168 Squadron and made responsible for getting the mountains of wartime mail to Canadian troops in Europe and the Middle East.

Even when the war had ended, the Mailcan B-17s continued their runs, and between October 19 and November 16, 1945, 168 Squadron flew Canadian Red Cross supplies from Rockliffe to avert famine in Poland. During the operation one B-17 crashed at Munster, Germany, killing the five crew members. On April 21, 1946, when 168 Squadron was disbanded at Rockliffe, its B-17s had flown 240 transatlantic flights and been gradually replaced by Liberators. The Flying Fortresses had worn many guises since their inception as Model 299 in 1934, but none as cheerful as the Mailcan flights.

Boulton Paul Defiant.

BOULTON PAUL DEFIANT

When German auto-maker BMW bought the Rolls Royce motor car company in 1998, they sent their young executives around to the British factory to try to figure out how the world's most finely crafted automobiles were built. After the tour one of them asked what the model of the Spitfire was doing in the corporate dining room. "That my boy," he was told, "is what your Luftwaffe chaps saw in their rear-view mirrors." It has been said that if the Battle of Waterloo was won on the playing fields of Eton, the Battle of Britain was won at the Rolls Royce factory where they made the Merlin engine.

The Merlin may have powered the Spitfire, Hurricane, and Lancaster to victory, but not even it could save the Boulton Paul Defiant. Here it was a case of a thoroughbred racehorse pulling a cart. In 1935 the Air Ministry in all its wisdom called for a two-seater fighter with a radical concept: all of its armament would be in a dorsal turret. The Boulton Paul Aircraft company had just built the Overstrand bomber and tendered a design that had a French hydraulic turret. It was two years before the Ministry approved it and ordered eighty-seven of the aircraft, now christened Defiants. It is easy today to dismiss the fighter with a gun turret saddled onto its back as a harebrained scheme that took the lives of most of its pilots, but when conceived, the Defiant was a radical innovation to aerial tactics. The aerial strategists of the early 1930s foresaw fleets of enemy bombers swarming over Britain. Defiants would manoeuvre into these formations, and their power-operated turrets, rather like Nelson's triple-deck ships of the line, would fire broadsides into the bombers. Thus the Defiant had no forward-firing guns; it was meant to be a bomber-killer, not a fighter aircraft. The logic was that by having his own gunner, the pilot was free to concentrate on flying the plane. Not to be outdone, the Royal Navy also wanted its own power turret fighter, and the Blackburn company obliged with the Roc, a naval version of the Defiant, which Boulton Paul also built.

Due to the lengthy testing of the aircraft after the turret was added, the first Defiant would not fly until July 30, 1939 — four years after the Air Ministry bid had been issued. It was powered by the same Merlin engine as the Hurricane, but the turret slowed it down considerably and the drag made it less manoeuvrable than the fighter. In the Battle of France, the Defiants initially ambushed Luftwaffe fighter pilots who, thinking they were Hurricanes, attacked from the rear, only to find themselves face to face with four .303 Browning machine guns. But once the element of surprise wore off, the German pilots realized that by meeting the Defiants head on, they were easy prey. Never designed to dogfight with Me109s, the Defiant now found itself used as a fighter over the beaches at Dunkirk and suffered appalling losses. On May 13, 1940, of the six Defiants that 264 squadron sent against Me 109s, five were shot down. Boulton Paul tried to compensate by re-engineering the Mark II with the Merlin XX engine, a larger rudder, and an increased fuel tank, but the losses during the Battle of Britain forced the RAF, on August 28, 1940, to withdraw all Defiants from daylight flying. It seems that none of those 1930s strategists had thought that the enemy bombers would be accompanied by fighters.

The aircraft was reconfigured as a night fighter, painted black, and sent to equip thirteen squadrons (two of which were RCAF) in time for the Blitz; on September 15, 1940, it accounted for its first kill. A year later, fitted with radar, the new Defiants had some success until they too proved too slow even for the German night bombers and were relegated to target tug and air-sea rescue duties for the remainder of the war. The Rocs were similarly unsuitable and, fortunately for the RCAF, only a single example was sent to 110 squadron on November 12, 1940, for assessment — and rejected. The last Defiant flew in the RAF in 1947 and resides today in the RAF Museum in Hendon in the black night fighter finish.

HANDLEY PAGE HALIFAX

Not as famous as the Avro Lancaster (but better known than the Shorts Stirling), the Halifax is always regarded as the "other" British heavy bomber of the Second World War. Thought of as slower and less trustworthy than the Lancaster, the "Hallie" lived in the shadow of its glamorous cousin, becoming the Cinderella of the Allied bombing force.

Conceived by Handley Page, it was also made by four other manufacturers and served in a staggering variety of roles. From the day it entered service on October 11, 1940, to March 17, 1952, when it made its last RAF sortie, the Halifax carried bombs, mines, torpedoes, rockets, paratroops, freight, and passengers. It towed huge Hamilcar gliders for the Airborne Regiment from England not only to Norway, Normandy, and the Rhine but also across to North Africa for the invasion of Sicily. It dropped arms and agents to the resistance movements for the Special Operations squadrons. It was in service with Coastal Command, and with No. 8 (Pathfinder) Group. Not only did it hold the distinction of being the first RAF four-engined "heavy" to bomb Germany, but it was the only British heavy bomber in the Far East and Palestine, even bombing the Afrika Korps from Egypt.

Like the Lancaster, the Halifax was initially envisaged as a twin-engine bomber also employing the Rolls Royce Vulture engines. When these proved unsatisfactory, four Merlins were substituted; later Mark III versions used Bristol Hercules radial engines. Admittedly, the Halifax was cumbersome when compared with the Lancaster, Handley Page's experience in bombers having extended only to its unconventionally shaped Hampden bomber. But when the front turret was removed and Perspex nose substituted, the flame-damping exhaust "muffs" reduced, "barn

Handley Page Halifax of RCAF 6 Group.

Courtesy of the DND

door" rectangular fins added, and engine cowlings redesigned, later marks of Halifax had an aerodynamic cleanliness that improved bombing accuracy if not speed.

But its failings were legendary — it performed poorly and was soon assigned to diversionary raids, mine-laying, and less heavily defended targets. It was given loads that weighed a third of a Lancaster's, usually bulky incendiaries that, if lighter, were more inflammatory when hit. Its ceiling was 22,000 feet, higher than the Stirling but lower than the Lancaster, but it had a better cruising speed (225 miles per hour) and range with maximum fuel load (2,785 miles) than either aircraft. If it had a reputation as being the most likely in a raid to be "coned" by the German searchlights and hit by flak and radar-guided rockets, post-war statistics showed that of the three bombers, the Halifax was the easiest air-craft to escape from — of its seven crew members 29 percent escaped, compared with 23 percent from the Lancaster.

By far the greatest user of Halifax bombers was the RAF's No.4 Group, which operated twelve squadrons from 1941 onward, filling the skies above their home bases in Yorkshire. In 1943 they were joined by Halifaxes flown by RCAF aircrews. Leeming, Croft, Skipton-on-Swale, and Tholthorpe are barely visible today on a map of Yorkshire, but during the war, these little villages were the Allied "front line" against Hitler as the airfields between Middlesborough and York were home to 6 (RCAF) Group. Integrated into the RAF Bomber Command then under Sir Arthur Harris, 6 Group was Canada's major contribution to Allied victory and was commanded by Air Vice Marshal C.M. "Black Mike" McEwen, a First World War Canadian fighter ace. Coming from the home of the Commonwealth Air Training Plan, the Canadians were thought to be better trained than their Allied colleagues but (and this was a British opinion) less disciplined, too happy-go-lucky — and more popular with women.

The Canadian government hoped to equip the whole Group with Canadian-made Lancasters, but as late as 1944, eleven squadrons were still flying Halifaxes. The bombers from 428 squadron inaugurated high-level minelay-ing when, on January 4 and 5, 1944, they dropped mines by parachute from 14,000 feet into the inner harbour at Brest. If most RCAF Halifaxes had a gallery of "nose art" that verged on the mildly pornographic, the squadrons themselves were patriotically named after Canadian birds (Goose, Snowy Owl), animals (Moose, Porcupine), or cities (Leaside). Some had more obscure labels: if 425 was "Alouette" because it was primarily French Canadian, 428 was called the "Ghost" squadron because a few months after its formation none of the original crews had survived.

F/L William Rodney DFC and Bar, trained at No.4 ITS, No.16 EFTS and No.4 SFTS and was commis-sioned in 1942. He remembers the Halifax "with prejudice."

Our boss, Sir Arthur Harris, didn't like the Halifax. He preferred the Avro Lancaster, which, with its longer bomb bay, carried a greater load. Moreover, the Lancaster, essentially an overgrown Anson, was easier to fly. The Halifax, especially the earlier Marks (1, B1, B2, and B5) with their artistically shaped fins and rudders, required judicious use of throttles and brakes on takeoff before the rudders began to bite. Climb was moderate at best, and service ceiling with a full load was not easily attained. At height and during evasive manoeuvres the aircraft tended to be sluggish in its responses, and rudders were prone to over control if the flying was poorly executed. Once it was reconfigured with the useless front turret eliminated, the mid-upper gunner's camel-humped turret replaced with a more aerodynamically efficient version, the D-shaped fins and rudders installed, and more powerful Bristol Hercules XVI radial engines substituted for the Roll Royce X liquid-cooled power plants, the Mark 3 and 3b variants became efficient, pleasurable aircraft to fly.

With the increased power there was a tendency on takeoff for the Mark 3 to swing to starboard, but the effect was easily countered by throttle use and the new rudder's greatly increased effectiveness. The tail came up easily and at all up weight the craft could be lifted off at around 100 mph indicated airspeed (IAS). After undercarriage and flaps (if any were employed) were raised the Mk3 climbed rapidly at 140 IAS, outpacing the Lancaster dramatically until it neared service ceiling. However, to get the aeroplane near or at its all-up weight above 20,000 feet required patient coaxing. Changes in trim were one of the Mk3's constants. Raising the undercarriage caused it to nose up, while lifting flaps induced a tendency to nose down. Opening the bomb doors brought the nose up slightly. These characteristics were easily countered by the trim tabs, all of which were very responsive and powerful. The elevator tab was particularly effective. In dives at high speeds the aircraft became tail heavy, the rudders particularly so, and I used to trim into the manoeuvre to compensate for that effect. These mild eccentricities had an unheralded benefit. On operations the need for patient attention to trim had the virtue of not allowing one to dwell excessively upon the dangers at hand.

All Marks of the Halifax that I flew stalled gently, with no tendency to drop a wing. Slight aileron buffeting preceded stalls, providing ample warning. Recovery was effected easily by easing the control column forward. The Mark 3, especially the "B" version with its extended wing tips, flew very nicely at all levels, especially once trim was adjusted. At low level it was responsive, and for a heavy aircraft, surprisingly nimble. If weather conditions were right and visibility was reasonable I used to return from operations at low level — anything from one hundred to three or four hundred feet depending upon terrain and power lines — in order to minimize the possibility of night fighter attacks from below. At low airspeeds, below 130 to 150 IAS, the aircraft was easier and more comfortable to fly with 30 to 35 degrees of flap. For the landing approach 110-115 mph with power was recommended. I generally came in at 120 as a reinsurance policy. Despite its comparatively short fuselage achieving a good three pointer with the Halifax wasn't difficult; but it demanded attention to do so.

Perhaps the Halifax's most redeeming features were its robustness, and its crew dispositions. With the wireless operator, navigator, and bomb aimer positions located in the aircraft's forward section below the pilot's cockpit, and with the escape hatch directly below the navigator's station, the possibilities of abandoning a damaged aircraft were, as Operational Research statistics confirm, roughly double those of a Lancaster. Emergency exits for gunners in both aircraft were comparable. In essence, the old "Hallybag" was a fine aeroplane, which, like the Hurricane, was never accorded its just recognition.

Handley Page Halifax.

Handley Page Halifax.

The last RCAF squadron to join 6 Group was 415 "Swordfish" squadron, until then operating Albacores in Coastal Command. In February 1944, Minister for National Defence Charles G. Power asked that 415 be transferred immediately to Bomber Command, but the British refused to release it until after the D-Day landings. The entries in the squadron's records book for the first few months that it was a bomber squadron give a snapshot of events that today seem implausible. While the transfer from Thorney Island to the airfield at East Moor, Yorkshire, was to take place July 12, 1944, no one told the squadron until that very day — with the first operational bomber sortie scheduled in two weeks time. Converting from Albacore biplanes to Halifax Mark III heavy bombers was difficult enough, yet incredibly, on July 26, sixteen crews climbed into their "Hallies" for their first bombing raid over Hamburg. It wasn't to be an easy first mission. Sadly, the first Halifax on takeoff, "U for Uncle," lost power on the starboard outer engine before getting airborne. It swung off the runway, broke in half, and caught fire. The other fifteen Halifaxes went by it, everyone concerned about their friends in the burning aircraft. One Halifax did not return from the raid. This baptism of fire proved to be a dress rehearsal for what came next. On August 21 two of the squadron's aircraft collided over Selby, Yorkshire, killing twenty-five men — twelve were aircrew, including the commanding officer, and the other thirteen were groundcrew. The squadron record book states that when the new commanding officer, Wing Commander J.H.L. (Joe) Lecomte arrived, "He made a deep impression by dealing with the mountain of paperwork (from the transfer) awaiting him. He actioned it all straight into the potbellied stove." A native of St. Theodore d'Acton, Quebec, Lecomte outlined three priorities for 415: that they flew on operations as much as possible, that when they weren't doing that they were on flying training, that when the first two could not be performed, they should have a party. The Canadians settled into the pattern of bombing raids that the other 6 Group squadrons had endured since 1943. Typical was the November 2, 1944, Dusseldorf raid, when of the seventeen Halifaxes sent out, two failed to return.

The Canadian squadrons never managed to convert fully to Lancasters and finished the war with six squadrons of Halifaxes. Among them was 415, which had kept the "Swordfish" insignia — fortunately, as they are today a maritime patrol squadron operating CP 140 Auroras. Among the Allies, the RCAF bomber crews had the second highest rate of casualties (at 17.8 percent) after the RAF, which incredibly lost 70 percent of its aircrew. After the war Halifaxes were converted to civil uses, carrying passengers as Haltons. So many were available that several were bought by the small charter air companies (like Eagle Aviation) that sprang up immediately after the war; some even flew on the Berlin Airlift. An example of how low a priority the preservation of a Halifax was can be inferred from the fact that of the 6,176 built, not one was saved for posterity. There are two Halifax aircraft preserved today. The one at the Yorkshire Aviation Museum in Elvington, Yorkshire, is a reconstruction from a Halifax that crashed and had been used for decades by a farmer as a chicken coop.

The other is at the RCAF Memorial Museum in Trenton, Ontario. On September 5, 1995, Halifax NA337 was brought to the surface of Lake Mjosa in Norway. Although this aircraft was not deployed to a Canadian squadron it represents a memorial to the many Canadians who flew the type. "X for X-ray" was part of RAF 644 squadron and was carrying out a supply drop mission on the night of April 23, 1945, to the Norwegian Resistance. Following the successful delivery of the cargo they came under German A-A fire defending a

Courtesy of Scott Knox

Halifax wreck found in Belgium: engine exhaust, carburetor, and magneto.

rail bridge. The starboard inner engine caught fire and the outer engine had also been hit. The captain, Flight Lieutenant A. Turnbull, DFC, had no choice but to ditch in Lake Mjosa. After initially surviving a rough ditching all perished but one, tail gunner Flight Sergeant T.H. Weightman, the only one able to make his way to the life dinghy. NA337 now lay in 235 metres (770 feet) of water. In February 1994, a representative from the Halifax Aircraft Association contacted the Norwegians who knew the location of the aircraft and had sonar pictures of it. After receiving salvage rights, permissions, and funding the project began. Initial design of the lifting rig was undertaken by Knox Technologies. The rig was to be inserted under the leading edge on either side of the fuselage between the wing root and inner nacelles. Careful attention had to be given to the centre of gravity of the submerged aircraft; if the design of the lifting rig was not accurate, the aircraft might slip out of the rig during the lift to the surface. The rig and aircraft were lifted by a winch and cable mounted on a barge above.

Once hoisted to just below the surface of the lake, the barge slowly brought the Halifax to where she could be dragged up on shore. The sled that was to be used failed under the strain, and eventually NA337 was lifted on shore by two cranes. CAF personnel treated NA337 for corrosion and disassembled and crated her for shipment to her new home at the RCAF Memorial Museum at Trenton, Ontario. Also recovered are parts of Halifax LW682 from No. 6 Group (RCAF) 426 Squadron shot down over Belgium on May 13, 1944. The restoration of a whole Halifax (in static condition) from both these wrecks would be a fitting memorial to a great aircraft.

Catalina/Canso.

Courtesy of the DND

CATALINA/CANSO

Winston Churchill admitted that the only thing that really frightened him during the war was the U-boat peril. It did not take, he said, the form of flaring battles and glittering achievements, it manifested itself through statistics, diagrams, and curves unknown to the nation, incomprehensible to the public. He might have added that the battle was partially won through endless patrols by Catalinas and Cansos. In their twenty-year service with the RCAF in war and peace, these ungainly-looking aircraft became the battle steeds for many Canadian heroes.

If, for the sake of argument, historians were to choose the single aircraft that changed the course of the Second World War, it might be the DC-3 or the long-range Mustang or perhaps the B-29 that dropped the atomic bomb. The Catalina/Canso flying boat would definitely not come to mind. For as Churchill realized, its successes were lonely ones, its battles far from land and the people it protected. Ungainly, without armour or self-sealing tanks, and, unlike the Shorts Sunderland, poorly armed, the Catalina had two saving graces: its maximum range was 2,500 miles and it cruised at a mere 117 miles per hour. It was the crew of an RAF Coastal Command Catalina that on May 26, 1941, found the mighty German battleship *Bismarck*, allowing Admiral Somerville's Swordfish torpedo bombers a chance to cripple it while the Royal Navy's destroyers closed in. The mainstay of American long-range reconnaissance in the Pacific, the PBYs became the early warning system for Admiral Chester Nimitiz, again and again locating the Japanese carriers before the Battles of Midway and Guadalcanal.

Originally built by Consolidated-Vultee at San Diego, California, to the Americans she was the PBY (PB for Patrol Bomber and Y designating the manufacturer, Consolidated). Consolidated had experience in long-range over-water flying with its P2Y-1 model in 1933 and 1934, which had broken records for non-stop flights from San Francisco to Pearl Harbor, a distance of 2,399 miles. It was the British who named the PBY after the island off the California coast. The Canadian version was an amphibian named Canso, after the Nova Scotia landmark, and was developed from the Consolidated XPBY-1 amphibian, first flown on May 19, 1936. While many air forces, from the Norwegians to the Australians, flew them, after the RAF, it was the RCAF that used the most Catalina/Cansos. The RCAF had thirteen squadrons — five on the Pacific coast, six on the Atlantic coast, and two over-

Catalina/Canso.

seas. Wresting them away from the RAF's allocation took some heavy influence. The first ten Catalinas to operate with Eastern Air Command's 116 Squadron at Dartmouth in 1941 were requested by Air Commodore A.E. Godfrey himself, who saw that the RCAF Digbys and Stranraers did not possess the range to cope with the U-boats.

Until the advent of the VLR Liberator, the Catalina was the only aircraft able to protect the convoys from the German wolf packs, and sometimes at maximum range. The need for long-range convoy protection to keep the Atlantic lifeline open was so critical that in 1941, Canadian Vickers Ltd. in Montreal and Boeing Aircraft of Canada in Vancouver were contracted to build the PBY-5 and PBY-5A. In record time, by early 1942 the first Canadian Vickers Cansos were ready, the first thirty-six aircraft going to the RAF to replace the Catalinas that the RCAF had received earlier from them. The first of 240 Boeing PBY-5A Cansos was completed on April 3, 1943; they were sup-plied entirely to other air forces — 193 to the RAF, 41 to the RNZAF, and 6 to the RAAF.

Of the many thousands of hours of Canso antisubmarine patrols, only a couple examples are needed to give testimony to the courage and tenacity of the men who flew them. On May 4, 1943, a battered convoy encircled by a U-boat wolf pack was limping towards St. John's, when a Canso from RCAF No. 5 squadron depth-charged and sank U-630. It was all the more remarkable because it did so 650 miles from its base at Torbay. The pilot, Second Lieutenant B.H. Moffitt, had achieved complete surprise as the U-boat was on the surface, its commander not expecting a land-based aircraft so far out. Three hours later and in the same area, a second RCAF Canso, flown by Flight Lieutenant J.W.C. Langmuir, surprised the fully surfaced U-438 and straddled it with depth charges. This one chose to fight it

Courtesy of Canadair

In 1941, Canadian Vickers Ltd. Montreal (shown here) and Boeing Aircraft of Canada, in Vancouver, British Columbia, were contracted to build the PBY-5 and PBY-5A.

Courtesy of the National Archives of Canada

Surrender of the German submarine U-889. The aircraft is a Consolidated Canso A flying boat of No. 161 Squadron, RCAF. Shelburne, Nova Scotia (vicinity). May 1945.

out, keeping its guns trained on the Canso. Langmuir stayed out of its range until HMS *Pelican* could be detached from the convoy to sink it. For their actions that day, both Moffitt and Langmuir were awarded the Distinguished Flying Cross. The events demonstrated the Catalina's advantage and its Achilles heel: its inexhaustible range and its peashooter armament of .303 Brownings against 20-mm and 37-mm German cannons.

The RAF supplied Catalinas to the two RCAF squadrons overseas, Nos. 413 and 162. Formed on Dominion Day 1941, 413 was first posted to Sullom Voe in the Shetland Islands to protect the Murmansk-bound convoys. It wasn't a salubrious posting for man or machine, and typical were the conditions on November 11: a severe storm sank four of the Catalinas at their moorings. In early 1942, when the Japanese fleet threatened to capture Ceylon, 413 moved to Koggala on that island to serve in an early warning role. Hardly had he flown the first RCAF Catalina to Koggala when on April 5, 1942, Second Lieutenant L.J. Birchall sighted the Japanese fleet approaching the British colony. His radio warning was intercepted and the Catalina was shot down. Birchall endured three years of captivity as a POW in Yokohama, Japan. But the warning he had given enabled the British to scatter their fleet out of the harbour, preventing another Pearl Harbor-type attack.

Based at Reykjavik, Iceland, three Catalinas from 162 (BR) Squadron were detached to Wick, Scotland, preceding the D-Day landings. On June 13, 1944, Wing Commander C.G.W. Chapman would sink U-715 north of the Shetlands, a signal victory in desperate circumstances. For another 162 pilot, battle would prove fatal: Flight Lieutenant David Ernest Hornell and his six men took off in Catalina 9754, code-named P, from Wick at 9:30 A.M. on June 24 for an eleven-hour patrol. Ten hours later, the starboard gunner sighted a fully surfaced U-boat (U-1225) five miles away travelling at high speed on the port beam. Hornell turned at once to attack. The official citation takes up the story:

The U-boat altered course. The aircraft had been seen and there could be no surprise. The U-boat opened up with anti-aircraft fire which became increasingly fierce and accurate. At a range of 1,200 yards, the two front guns of the aircraft replied; then its starboard gun jammed, leaving only one gun effective. Hits were obtained on and around the conning tower of the U-boat, but the aircraft was itself hit, two large holes appearing in the starboard wing. Ignoring the enemy's fire, Flight Lieutenant Hornell carefully manoeuvred for the attack. Oil was pouring from his starboard engine which was by this time, on fire, as was the starboard wing, and the petrol tanks were endangered. Meanwhile, the aircraft was hit again and again by the U-boat's guns. Holed in many places, it was vibrating violently and difficult to control. Nevertheless, the captain decided to press home his attack, knowing that with every moment the chances of escape for him and his gallant crew would grow more slender. He brought his aircraft down very low and released his depth charges in a perfect straddle. The bows of the U-boat were lifted out of the water; it sank and the crew were seen in the sea. Flight Lieutenant Hornell contrived, by superhuman efforts at the controls, to gain a little height. The fire in the starboard wing had grown more intense and the vibration had increased, culminating in the burning engine falling off. The plight of the aircraft and crew was now desperate. With the utmost coolness, the captain took his aircraft into wind and, despite the manifold dangers, brought it safely down on the heavy swell. Badly damaged and blazing furiously, the aircraft settled rapidly.

Then came a twenty-one-hour ordeal for the survivors in a partially capsized dinghy, during which Hornell, by his cheerfulness and leadership, kept their spirits up until a Catalina of the Royal Norwegian Air Force spotted them. He died shortly after being rescued and is buried in the Shetland Islands. On December 12, 1944, at Government House in Ottawa, his widow, Genevieve Hornell, accepted a simple bronze medal on behalf of her husband from the Earl of Athlone, the Governor General. David Hornell had become the first member of the RCAF to receive the Victoria Cross.

The battle of the North Atlantic continued to the very last day of the war, ending only at midnight on May 8, 1945, when under the orders of Gross Admiral Doenitz the remaining U-boats were ordered to surface and surrender at the nearest Allied port. For the air force crews, it was the first opportunity they had to look at the enemy they had hunted for so long. Of the 1,162 U-boats commissioned in the German Navy, 727 were destroyed by enemy action, and of this number 288 were sunk in open water by aircraft operating alone.

When Canso/Catalina production ceased on May 19, 1945, Canadian Vickers had built 369 aircraft of this type. The RCAF received 139 aircraft from Vickers and the USAAF took delivery of 230, designated GA-10A. The last Catalina built by the Boeing Aircraft Co. was christened "David Hornell VC" by Air Vice Marshall F.V. Heakes, the officer commanding Western Air Command in 1945. The ceremony acknowledged a gallant pilot and aircraft.

The shortage of civil airliners immediately after the war prompted Canadian Pacific Airlines to buy four ex-RCAF Cansos (CF-CRP, -CRQ, -CRR, and-CRV) for its regional network. As there was no runway at Sandspit, the amphibious aircraft were reconditioned into civilian guise, renamed Landseairs, and put on the Vancouver–Prince Rupert–Sandspit route. CF-CRR had, while in the RCAF, sunk the German submarine U-342 on April 17, 1944. There are few airlines that can claim such a distinction. Two of the Cansos were written off in accidents. On May 11, 1953, CF-CRV hit a submerged log while landing at Prince Rupert and flipped over, killing the stewardess and a passenger. The last to go was the battle-honoured CF-CRR, which was sold to Northland Airlines in 1960.

Post-war, the RCAF used Cansos in photographic squadrons and Rescue Units, where they flew into hazardous terrain and treacherous weather conditions. The number of occasions warrants a book, but here too a couple will suffice. In the late summer of 1947, in the Canadian exploratory polar mission POLCO, a Canso from 413 Photographic squadron became the first RCAF flying boat to be based among the barren islands surrounding the North Pole. For his skill, resourcefulness, and courage under trying conditions, its pilot, Flight Lieutenant John Francis Drake, was awarded the Air Force Cross. On June 18 and 19, 1950, Flight Lieutenant Lawrence Bell Pearson, flying a Canso of No. 103 Rescue Unit from Greenwood, Nova Scotia, was awarded the Air Force Cross for a mercy mission at Ungava Bay.

The last Canso was withdrawn from RCAF service on April 6, 1962. Indisputably the vehicle of heroes, an amphibious angel during and after the Second World War — the Catalina/Canso.

Hawker Typhoon.

HAWKER TYPHOON

The Typhoon, like the Hurricane, Tempest, and Fury fighters, was the work of one man, Sydney Camm, the aeronautical genius who would one day design the prototype to the Harrier. As soon as the first Hurricane flew in 1937, the Air Ministry wanted its successor, asking for a more powerful fighter with twice the firepower that could fly high enough to intercept bombers. Camm did manage to get the first Typhoon airborne on February 24, 1940, and might have put many more in production except that Lord Beaverbrook, the Minister of Aircraft Production, preparing for the Battle of Britain, ordered that all resources be concentrated on building only Spitfires and Hurricanes.

When it did make its debut, it looked as though even Camm could make mistakes. Through its early years, the Typhoon was unpopular with whoever saw, serviced, or flew it. With its breadboard-type wings and enormous chin radiator, it looked heavy and cumbersome compared with the svelte Spitfire, and unlike the tried and tested Merlin engine, its Napier Sabre was an unknown quantity. On its very first flight the aircraft almost fell apart and the test pilot was awarded the George Cross for bringing it down in one piece. When it became operational the RAF groundcrew were loath to deal with this seven-ton brute, and the forty-eight spark plugs in its monster engine proved difficult to start in cold, rainy weather, and even harder to keep synchronized. Scorn for Camm's flying elephant turned to animosity when it was discovered that the Sabre's cylinder sleeves distorted within twenty hours of use, the pistons broke through, engine oil spattered over the windshield, and it seized. And there was worse to come — the aircraft's thick wing disintegrated at high speeds, its tail kept falling off without warning (killing twenty-eight Typhoon pilots), and it couldn't perform above 20,000 feet. Cross-Channel flights were invitations to disaster; with so heavy an engine, no one expected to survive a ditching at sea. The pilots themselves abhorred the Typhoon, and for good reason. Carbon monoxide from the engine seeped through the firewall bulkhead, so they were instructed to put their oxygen masks on as soon as the engine was started. The cockpit, a heavy pillared affair, afforded little rear visibility and was entered by a little door, better suited, they said, to a motor-car, and impossible to bail out of.

This melancholy state of affairs was exacerbated when the heavy-chinned fighter was sent to the front-line squadrons on the British coast, every one of which soon had their own Typhoon gliders — if the pilots could bring

Hawker Typhoon.

them down successfully. Set to re-equip with Typhoons, squadron commanders protested and asked if they could keep their worn Spitfires instead. Worst of all, Typhoons were mistaken for the new German Focke Wulf 190, and several were mistakenly shot down by RAF and US pilots. Repainting them in a glaring colour scheme to aid recognition only made them uglier.

The initial engine, airframe, and cockpit woes were gradually corrected. The Bristol engine company adapted its Taurus engine sleeves to the Sabre and the bubble canopy was substituted for the Austin Seven door. Neither improvement helped its reputation, and Typhoon production was almost cancelled by the Air Ministry, led (it was said) by the Spitfire lobby. What saved the aircraft was the enemy. The appearance of the FW 190 was a compelling reminder to the Allies that the Luftwaffe was not defeated. These radial-engine fighters outflew the Spitfires, especially on low-level raids on the south coast. These nuisance raids affected the morale of the civilian population, and it was then that the Typhoons came into their own. Using their power at low levels, they could intercept and shoot the 190s down. Now Camm's design earned praise: it showed that the Tiffie could take the stress of low-level flying — pilots flew so low that the huge radiator scoops sometimes came back with branches and ships rigging in them. The cancellation order was rescinded and the Typhoon was reborn as a tactical fighter.

By 1943, with the Luftwaffe all but gone from British skies, the Spitfires were being misused on rhubarbs, low-level fighter sweeps over France. Designed for high-level combat, the Spitfire lacked the firepower and was vulnerable to ground fire. The chunky Typhoon was now recognized as a potent weapon, as its four Hispano cannon could fire 650 shells a minute. It was set free to rampage across France, destroying German military equipment, trains and shipping, V1 rocket sites, coastal batteries, and fortifications. It was said that their low flying gave great encouragement to the occupied population. Now the thick wing and four Hispano cannon could be used to good effect. It could not only take heavy punishment but dish it out — to shipping off the Dutch coast, flak batteries guarding fighter bases at Abbeville, steel plants, and trains. Typhoons were then adapted to be dive bombers (called Bomphoons), flying rhubarbs both day and night, sometimes at levels of two hundred feet between trees and farmhouses to escape enemy fire. Made heavier with amour plating, it was no longer considered an aircraft at all but a flying artillery piece, much like the A-10 Warthog gunship four decades later.

It was only a matter of time before someone realized that the solid, basic design that Camm had given the Typhoon could carry even more potent armament. In addition to the Hispano cannon, eight and later twelve rockets, each weighing sixty pounds, were fitted onto rails under the wings. Typhoon squadrons now sought out important, precise targets, relying on surprise as much as their high-explosive rockets. On the eve of the D-Day landings, aircraft of RAF 193 squadron hit the château where General Irwin Rommel had his headquarters. Killing the mastermind behind the coastal defences was a priority for the Allies. They demolished the room he should have been in but missed the general. But a few days later, Typhoons surprised him in his staff car in the open and wounded Rommel sufficiently to prevent him from taking part in the coming battle.

Gestapo prisons and German high command sites also became favourite targets for the rocket-firing Typhoons. But they were best known for the standing patrols they flew just behind the front line, waiting in cab ranks, ready to be called down by field commanders to provide target saturation in minutes. American general George Patton, no Anglophile, once said that the Typhoons were the best thing the Limeys ever built, that his armoured breakthroughs owed much of their success to calling down the cab ranks.

Courtesy of the DND

Hawker Typhoon.

Operating from airfields in Normandy was harder for Typhoons in the Tactical Air Force — the quartz dust on the fields clogged up their filters. When two RCAF aircraft burst into flames and crashed without reason, a Canadian wing threatened not to fly their aircraft. It was later discovered from one of the crash sites that the fuel vents of the Typhoon had become clogged when the Wing was operating low over the Normandy beaches, and caused fuel to flood the gun bays, creating flashbacks when the guns were fired.

A total of 3,330 Typhoons were built. Almost all were by Gloster, as Hawker was already at work in 1941 designing the Tempest, which Camm hoped would be a Typhoon with the bugs out. Besides the RAF, the air forces of five other nations flew Typhoons — Belgium, New Zealand, Norway, Poland, and Canada. In its success, Sydney Camm had been vindicated.

DE HAVILLAND 98 MOSQUITO

Salisbury Hall near London Colney is where Charles II cavorted with his mistress Nell Gwynne, scandalizing British society. Whatever they did, it wasn't half as radical as what would be designed centuries later just outside the manor house. For it was in a hangar built on the Hall's cabbage patch that the Mosquito was conceived.

The first impression of a Mosquito was its simplicity and beauty — a single line that flowed uninterrupted from nose to tail, from engine cowlings to the treaded tires of the undercarriage. The aircraft always seemed to sit up, looking pert and seductive — a pleasing shape that one could easily form a love affair with. Unlike the Beaufighter with its pugnacity, or the vicious warlikeness of the Typhoon, a Mosquito had a look of anticipation about it, as if it were about to win an air race for you. It was easy to love.

But for all its fame and success the Mosquito was an accidental aircraft. A wooden fighter? With the lumber and furniture industry sidelined by the war, it made sense, but would the Air Ministry be convinced? Lord Beaverbrook, the Canadian-born Minister for Aircraft Production, wanted de Havilland to drop the idea immediately and concentrate on building more Tiger Moths to train pilots. He only came around when told that the new aircraft used no strategic materials. At a time when air force thinking held that bombers had to be heavily defended by gunners (the B-17 had five gunners and yet sustained heavy losses), the concept of one relying on speed alone for its defence was too radical. Happily, those who had faith in the Mosquito were the same group of men at de Havilland who had built the DH 88 Comet racer for the MacRobertson race in 1934. Learning from this and the wooden construction of the DH 91 Albatross airliner, de Havilland's R.E. Bishop combined the virtues of the two into the Mosquito project.

DH Mosquito.

Courtesy of the DND

Purely from the economic point of view, coming as it did in 1940, it was cheap to build and amazingly adaptable. A bomber, night fighter, convoy escort, high-speed transport to Sweden, anti-shipping fighter, V-1 doodlebug destroyer, photo-reconnaissance, and Pathfinder marker — the Wooden Wonder was all of those. Glue and screw construction and twin engines notwithstanding, the Mosquito could take anything on board — cannon, rockets, 4,000-pound cookie bombs, or radar, either in the bulging H2S underneath or in the distinctive nose thimble. The wartime American media could never get over the fact that the Mosquito could carry as heavy a bomb load as a B-17 Flying Fortress — and deliver it at 400 miles per hour!

The first production batch was ordered by the Air Ministry on March 1, 1940, and the two-seater prototype was powered by two 1,300-horsepower Rolls Royce Merlins with an armament of four .303 Browning machine guns in the nose and four 20-mm Hispano cannon in the belly of the fuselage. The prototype, EO234, painted yellow for identification, was rolled out at Hatfield on November 21, 1940. The son of the company's founder, Geoffrey de Havilland Jr., personally test-flew it — recreating the flight for a movie years later — and even delivered some of the first few to operational squadrons.

In December 1940, C.D. Howe went to Britain (and was torpedoed along the way) to discuss Canadian wartime production, and on December 29 he saw the Mosquito prototype fly. He took the impression back to Canada,

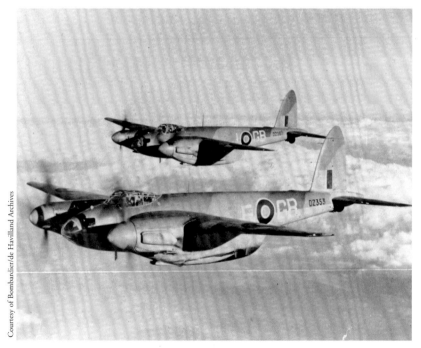

Courtesy of Bombardier/de Havilland Archives

and on July 7, 1941, Ottawa decided that the de Havilland plant at Downsview would build the aircraft, using Merlin engines built at the Packard auto factory in Detroit. Two senior de Havilland staff, Harry Povey and W.D. Hunter, were sent to Canada to adapt the design to Canadian materials. General Motors at Oshawa made the fuselages, Massey Harris the wings, Boeing's Canadian subsidiary at Vancouver the tail planes, and the Canadian Power Co. the flaps. It all came together at Downsview.

With the Blitz on, the Mark II was initially used as a home defence fighter equipped with the early radar sets, but its success led de Havilland to develop the Mark VI as a fighter-bomber or intruder for making offensive sweeps across the Channel. The prototype Mark VI flew

Using specially designed 200-gallon fuel tanks, 646 DH Mosquitos were ferried across the frigid Atlantic, with the loss of only 28.

on June 1, 1942, and production would eventually total 2,305. It was this version that became the most popular, at home and overseas. One of the uses that the Mark VI was put to was to be a clandestine airliner flying between Leuchars (in Scotland) and Stockholm. Lying on a mattress and Mae West (a slang term for a life jacket) in the bomb bay, several VIPs, including Sir Kenneth Clark, Sir Malcolm Sargent, and Professor Niels Bohr, took the BOAC Mosquito flight from Sweden.

By 1942, the new centimetric radar became available and the Mosquito was adapted to take the radar dish in a thimble nose radar dome, in place of the four nose machine guns. These were the Marks XII, XIII, XVII, and XIX. To combat the high-altitude Luftwaffe fighters, de Havilland was asked to produce a high-altitude Mosquito with a pressurized cabin. It did so in a commendably short period, and this was to be the Mark XV, with 1,680-horsepower Merlins that had two-speed, two-stage superchargers.

DH Mosquito.

In Canada, the first locally built Mosquito Mark VII, KB300, made its test flight on September 24, 1942, flown by G.R. Spardbrow and F.H. Burrell. The de Havilland panache and glamour blessed the project. Movie star Olivia de Havilland, a cousin of Geoffrey's, toured the plant, and Geoffrey de Havilland Jr., on a publicity visit to North America, took a Mosquito up for a series of aerobatic flights over Downsview. By the end of 1943, ninety-two Mosquitos had been completed in Canada, and, with the exception of the first fifty for local Operational Training Units, they were being flown across the Atlantic. The first operational sortie by a Canadian-built Mosquito took place on December 2, 1943, when BXX-KB161 bombed Berlin. Flown by Flight Lieutenant G. Salter and Wing Officer A.C. Pearson, DFM, of 139 Squadron, it was called New Glasgow after the Nova Scotia town. Thereafter Canadian-built Mosquitos played an increasingly important part in Bomber Command's raids.

The final wartime Mosquitos that saw extensive use were the anti-shipping versions, which had a Molins anti-tank gun of 57-mm calibre fitted to the nose, in addition to extra fuel tanks and armour around the cockpit to combat the heavily armed U-boats. It had an effective range of over a mile, thus outranged the 37-mm guns fitted onto the German submarines. The first two of these Mosquitos were sent to No. 248 Squadron at Predannack, Cornwall, in October 1943. Until then, surfaced U-boats close to their French bases had been able to rely on the protection of Luftwaffe fighters to come home, but the nimble little Mosquitos changed that. The Molins-equipped aircraft was first used on November 7, 1943, when Flight Officer A. Bonnet, RCAF, surprised the returning U-123 on the surface within sight of her base at Lorient. The Canadian pilot's armour-piercing rounds smashed into its conning tower, killing the petty officer, wounding two sailors, and, with a hole seven by two and a half inches, rendering the submarine incapable of diving. Though Captain Oberleutnant von Schroeter was able to make his base before the Mosquito was sent off by intense ground fire, it was a salutary lesson for the U-boats.

By the time the war ended, the Mosquitos remaining in Canada were declared surplus to the requirements of the RCAF, which chose the Mustang for home defence instead. In 1947, the Chinese Nationalist Air Force bought 250 of these for $10,000 each. They were shipped to Shanghai, where a factory was set up to assemble them under a supervisor from the de Havilland Downsview plant. To train the Chinese pilots, nine Mosquitos were kept at Downsview and, to keep to Canada's neutrality in the civil war, repainted in the markings of the Chinese National Aviation Corp — the country's airline. When the Nationalists fled to Taiwan, some of the Canadian Mosquitos found their way there.

In RAF service the Mosquito was finally replaced by another bomber that relied on speed rather than guns — the Canberra. The Wooden Wonder made its final RAF sortie on December 15, 1955.

Easily the best tribute to the aircraft came from Reichsmarshall Hermann Goering, who said:

It makes me furious when I see a Mosquito. I turn green and yellow with envy. The British who can afford aluminum better than we can, knock together a beautiful wooden aircraft that every piano factory over there is building and give it the speed which they have now increased yet again. There is nothing the British do not have. They have the geniuses and we have the nincompoops. After the war's over I'm going to buy a British radio set — then at least I'll own something that has always worked!

A former air ace and connoisseur of fine things, Goering knew a great aircraft when he saw one.

CONSOLIDATED B-24 LIBERATOR

The story is that Mrs. Reuben Fleet, the wife of the aircraft manufacturer's president, named it the Liberator. When her husband asked her to suggest a name for the new bomber, Mrs. Fleet discussed it with Edith Brocklebank, her children's governess. Edith was British and, perhaps in sympathy for all those countries in Europe under the Nazi boot, she suggested Liberator. For an aircraft that saw service in every theatre of the war and was built in greater numbers than any other, it was an appropriate name.

Yet in the popular imagination the Liberator was eclipsed throughout the war by the B-17 Flying Fortress — perhaps because it was too new and did not even exist until the war had begun. Consolidated Vultee had just moved from Buffalo, New York, to San Diego, California, and was about to build a flying boat that the French government was interested in. The chief engineers, Isaac M. Laddon and David R. Davis, had a winning design — in particular Davis had conceived a very slender wing that reduced drag and increased the range of the aircraft. When encouraged by the United States Army Air Corps, in January 1939 they switched to a bomber, changing the fuselage and adding a tricycle undercarriage instead.

Unlike the comely Flying Fortress, the Liberator's design was function over form: large oval rudders, a deep body with a catwalk that allowed the crew to access any position they were needed in, and four heavy 1,100-horsepower Pratt & Whitney R-1830 Twin Wasps engines. Even then it was realized that a single fin would be better and that the cockpit needed armour. At a speed of 273 miles per hour, it was sluggish, but its range was the selling point — 2,100 miles. The first aircraft flew on December 29, 1939, and almost immediately, the desperate French government placed an order

Consolidated Liberator.

for 175 of the export version, the LB 30A (LB for Land Bomber). France surrendered before the aircraft could be delivered and they were passed to the British as part of the Lend-Lease arrangement. The RAF had misgivings about using the Liberator as a heavy bomber, but they liked its inexhaustible range and adapted it for transport operations.

Initially, the Liberator came to Canada in sheep's clothing — as a civil airliner. When BOAC was charged with running the Return Ferry Service between Montreal, Canada, and Prestwick, Scotland, in May 1941, the British Air Ministry transferred six of the French LB-30As to it. With them BOAC began the first transatlantic air service between Canada and the United Kingdom. An added bonus was that Trans Canada Airlines at Dorval was asked to service the Liberators, especially their ice-prone Pratt & Whitney engines. Besides helping the war effort, the TCA employees realized the BOAC contract was preparation for the day when their airline would have its own multi-engine aircraft and also fly the Atlantic.

One of its first Liberator passengers was Prime Minister W.L. Mackenzie King. King had never flown (no Canadian prime minister had) but with the U-boat menace there was no other way for him to confer with British Prime Minister Winston Churchill. After a séance with his long-departed mother and former Prime Minister Wilfrid Laurier (also dead), he decided to cross the Atlantic by air. A BOAC Liberator was outfitted with a bunk and two reclining chairs for his use, and on August 19, 1941, he left St. Hubert in midmorning, touched down at Gander at sunset, and then slept soundly all the way to Prestwick. King returned by Liberator on September 7, becoming not only the first VIP to use the new airport at Dorval but also the first Canadian prime minister to fly. Unlike subsequent Canadian prime ministers, he never took to flying — nor to the Liberator. Winston Churchill was more adventurous and had a Liberator Mark II called "Commando" outfitted for his personal use.

The initial British distrust of the Liberator caused them to confine the aircraft to a transport role, and a single Coastal Command squadron, No. 120, was equipped with it to escort the Atlantic convoys. But because of the aircraft's range, it was the only one capable of providing air cover over the Atlantic Gap, the area between Newfoundland and Iceland, far beyond the range of other patrol aircraft and the killing ground for the

Consolidated B-24 Liberator aircraft carrying the Right Honourable W.L. Mackenzie King and party en route to England. Gander, Newfoundland, August 19, 1941.

Courtesy of the National Archives of Canada

U-boats. To bridge the Gap, the British asked the RCAF to extend their anti-submarine patrols to some eight hundred miles off Newfoundland, far beyond the range of 10 (BR) Squadron's Digbys at Gander. Liberators were the only aircraft capable of this, and by 1942 neither the Americans nor British were willing to give up them up. It must have irked many Canadians that at a time when their own coasts were in danger of U-boats, they could see these precious long-range bombers lined up at Dorval Airport to be ferried to distant parts of the British Empire.

Exchanged with the Indian Air Force for a Lysander, this former RAF Liberator was flown to Canada and is on display at the National Aviation Museum in Ottawa.

There were plants at San Diego, Fort Worth, and Tulsa mass producing Liberators; the Ford plant at Willow Run alone turned out twenty a month. Yet in 1943, there was a shortage of these aircraft. It took heavy shipping losses to the U-boats in the North Atlantic in the winter of 1943 before fifteen British Liberators were diverted to the RCAF. George Lothian and Lindy Rood, two pilots from TCA who had multi-engine experience, were called on by the RCAF to train their first Lib pilots. Two RCAF squadrons, Nos. 10 and 11, flew Liberators in an antisubmarine role, and when one sank the German submarine U-341 on September 19, 1943, it demonstrated that the Gap had been closed.

Overseas many RCAF officers flew Libs in an antisubmarine role; one of these was Flight Officer Kenneth Moore in 224 Squadron. On June 8, 1944, Moore, Warrant Officers Foster and McDowall, and Sergeant Hamer were pilot, wireless operator (air), navigator, and flight engineer respectively of a Leigh Light Liberator when their radar reported a contact twelve miles ahead. Moore sidestepped the aircraft to position the contact between the bright moon and himself and then turned towards it. McDowall caught sight of the U-boat against the line of the moon on the water, and Moore prepared the aircraft for the attack. Passing forty feet over the conning tower he released six depth charges: three fell to the starboard and three to the port — straddled! As they exploded, the rear gunner shouted, "Oh God, we've blown her clean out of the water." Returning later to the scene, they could make out oil and wreckage on the water — all that was left of U-629. Jokingly, Moore then said, "Now let's get another one." And they did. Ten minutes later, they attacked and sank U-373. For his actions that night, Flight Officer Moore was awarded the Distinguished Service Order. It had been a textbook depth charge attack — he had used neither his Leigh Light nor the homing torpedo.

Many older Montrealers will recall their city's use as the terminal for the ferrying of aircraft to the front lines and especially the day that it became a lot more real. A Liberator being ferried to the RAF in India took off from Dorval on the morning of April 25, 1944, in a light drizzle and under a low cloud base. Thousands of Montrealers heard it unable to gain altitude before crashing. The Lib missed the Dow Brewery but smashed into houses at the corner of Shannon and Ottawa streets, killing ten civilians and injuring many more.

There was a similarity with Britain's other major dominion. Because of the Japanese advances, Australia was cut off from Britain and its own military forces in the Middle East and Europe. In June 1944, using BOAC LB-30s, the Australian airline QANTAS began ten-hour flights from Perth (later Learmouth) to Colombo, Ceylon. Although the Japanese were not signatories to the Geneva Convention, to ensure its civilian status a kangaroo with "Kangaroo Service" was painted on the fuselage's metal finish.

By the war's end, the RCAF had taken 148 Liberators on strength, its 168 HT squadron using some for transport from Rockliffe, Ottawa. After the war, RCAF 426 Thunderbird squadron was re-equipped with Libs, exchanging them for Canadair North Stars in November 1947. The Liberator was by far the most numerous heavy bomber in the Second World War, with 18,188 built — compared with 12,730 B-17s.

North American Mustang.

NORTH AMERICAN MUSTANG

The P-51 Mustang gave truth to the fighter pilot's saying: "If it looks good, it flies good." Held to be the best all-round fighter to come out of the Second World War, the Mustang typified Anglo-American genius. The Merlin engine married to the North American body gave it excellent qualities: a long range, toughness, manoeuvrability, and firepower. If one aircraft profoundly affected the outcome of the Second World War, it was the P-51 Mustang.

This remarkable aircraft began its career as the North American NA-73, equipped with the Allison V-1710 engine. The British placed an order for 320 NA-73s on May 1, 1941, and no one knows who first called the aircraft the Mustang Mark I, but it entered RAF service in April 1942. It was a well-designed ship, making use of the technology of the laminar flow wing. Pilots discovered that their Mark Is were dependable and ran well, but the Allison engine made it a poor performer above 15,000 feet. As a result, they were assigned to RAF squadrons in the Middle East for ground attack, where their wide track undercarriage was appreciated. Then the Packard-built Rolls Royce Merlin with its two-speed, two-stage supercharger was substituted for the Allison and the results were beyond everyone's expectations. Heavier than its contemporaries (the Spitfire, Messerschmitt 109, and FW 190) it climbed more slowly, but at same engine settings the Mustang was thirty miles per hour faster in level flight. However, the replacement Merlin with its liquid coolant system made it highly vulnerable to ground fire — the coolant radiator under the fuselage centre was a prime target, and a single bullet through its lines would cause the Merlin to seize in seconds. But by now with the United States in the war, the Mustang was given another role, one that it excelled in: high-level fighter escort, the famous Little Friends of the B-17 bomber formations.

During the war, Canadians flew Mustangs with the RAF and in five RCAF squadrons overseas. One was Flight Lieutenant Duncan Marshall Grant, who, in 1941, joined 400 Squadron, an Army Co-operation unit at Odiham, England. Nicknamed Bitsy because he was short, Grant initially flew Lysanders and then Tomahawks. It was when 400 received the Mustangs that he found his specialty: train busting. His squadron set a record, demolishing troop-carrying trains in France, with Grant personally destroying some thirty locomotives. On April 13, 1943, he shot down his first aircraft, a Dornier 217, in the Paris area, and later he added another aircraft and a probable to his score. Grant was awarded the Distinguished Flying Cross for his fine fighting spirit and determination. But on August 28, 1943,while on an airstrike over Ault in northern France, his Mustang was hit by flak. Bitsy Grant died in the crash and is buried at the Canadian War Cemetery at Dieppe. In October 2002, his old unit, now 400 Tactical Helicopter Squadron at CFB Borden, recognized his spirit by naming Hangar 18 the Grant Building in his memory.

The Mustang also served in Korea with the Americans and South Africans and was given to over fifty other air forces. Late in the war, the RCAF requested an allocation to obtain Mustangs as replacements for the Kittyhawks and Hurricanes in the home defence squadrons. By the time the allocation

Personnel with North American Mustang IV aircraft of No. 442 (City of Vancouver) Squadron (Auxiliary), RCAF, at Sea Island, Vancouver, July 13, 1951.

North American Mustang.

was granted, the war had wound down and ended. There was no longer any need for the Mustangs in post-war RCAF fighter and reconnaissance squadrons as there was now a surplus of de Havilland Mosquitos, many of which were in storage. The military debated the merits and costs of using the existing Mosquitos versus buying the Mustangs, taking into account wooden versus metal airframes, airframe life, fuel consumption (the Mustang took less), aircrew, and engine maintenance (the Mosquito had two engines, the Mustang one, thus it was easier to maintain) ... finally settling on the Mustang.

In 1947, the RCAF purchased the first thirty P-51s to equip a Regular Force Fighter Reconnaissance squadron at CJATC Rivers. A further one hundred Mustangs were bought and delivered in 1950 and 1951 for use by the RCAF (Auxiliary) squadrons and to get a few of the reactivated Regular Force squadrons flying until their Sabres arrived. In all, six Auxiliary and two Regular Force squadrons flew them, managing to write off forty-one Mustangs in the process. They were also used in a variety of roles and units: Air Armament school, Pilot Weapons School, and at Chatham OTU towing targets for the Sabres.

All Mustangs were retired from RCAF operations in the late summer and early fall of 1956. Eighty-seven were sold to civilians in the United States and two were kept as display aircraft — one at the National Aviation Museum in Ottawa and the other originally at St. Jean, Quebec, but now at the Imperial War Museum in London.

Auster.

AUSTER AOP

The original uses of aircraft by the military were to observe the movements of the enemy and spot for the artillery. In both world wars small, unarmed aircraft fulfilled this role in all armies. With a vast light plane industry that built Stinsons, Cubs, and J-3s, when the United States entered the Second World War, its military had a variety of aircraft for this purpose. It was one of these, a sports plane built by Taylorcraft in 1938, that the Auster-Beagle Company of Britain chose to build under licence in 1944.

Made of composite wood, metal, and fabric covering, the Auster in Australian, British, and Canadian service used a 130-horsepower DH Gipsy Major engine. The two-seater was cheap to build and operate and had the perfect short takeoff and landing qualities necessary for the battlefield. In 1946, Auster-Beagle changed its name to Auster and all its aircraft were known as Austers. A total of sixteen hundred Auster Army Co-operation aircraft in Marks I to VI were built for use in the British Army Air Observation Post squadrons. The RAF maintained the little aircraft but the pilots and observers were from the Royal Artillery, combining the skills of flying and gunnery. Austers first flew in the RCAF with 664 Squadron when it was formed on December 9, 1944. The two other RCAF squadrons that flew the aircraft were 665 from January 22, 1945, and 666 from March 5, 1945.

The RAF, the RAAF, and the RCAF also operated Austers in Korea, the aircraft performing much the same duties as their First World War ancestors had: artillery spotting, photo reconnaissance, and light transport. An indication of what it must have been like to fly one in a battle zone is taken from the *Canada Gazette* of November 14, 1953. Vancouver born Captain Peter Joseph Tees, RCA, flew Austers in the 1st Commonwealth Division's AOP in

Korea from September 1952 through the winter of 1953. In recommending him for the Distinguished Flying Cross, the commanding officer of the Royal Artillery wrote:

> On one occasion he carried out a crash landing due to subzero temperatures encountered and on two others he returned to base after engine failures, making successful landings. He controlled the guns not only of the Division but also the Corps artillery in 185 sorties and conducted 453 shoots engaging enemy troop concentration, camouflaged guns, bunkers, groups of men and vehicles. In spite of the increase in density of hostile anti-aircraft guns he has ignored his own safety in order to obtain the best observation of his targets.

The Austers of three RCAF AOP squadrons were replaced in 1958 by Cessna L-19s (Bird Dogs). Like the Austers, they too were overhauled by MacDonald Brothers (Bristol Aerospace Ltd.) in Winnipeg. On July 1, 1972, the AOP ceased to exist as a fixed wing and was absorbed into the Air Force as helicopter squadrons. AOP 1RCHA, then in Germany, became 444 Tactical Helicopter Squadron, AOP 2 RCHA Gagetown became 403 OTU, AOP 3 Shilo became 408 Tactical Helicopter Squadron Nanaimo/Edmonton, AOP 4 RCHA Petawawa became 427 Tactical Helicopter Squadron, and AOP5 RALC Valcartier became 430 Escadrille, Valcartier.

Courtesy of the RCA Museum

A pilot-observer in an Auster: the original uses of aircraft by the military were to observe and "artillery spot," and in both world wars light, unarmed aircraft like the Auster fulfilled this role.

DE HAVILLAND VAMPIRE

Probably because they invented the language, the British have a talent for giving their aircraft evocative, soul-stirring names. Venom, Buccaneer, Spitfire, and Gladiator are names that are calculated to inspire their proponents as well as strike terror in the hearts of the enemy, even if the actual aircraft didn't. Naming otherwise fine aircraft Mitchells or Zeroes or Rumplers is doing an injustice to them. Yet when de Havilland built its DH 100 jet fighter, it was initially christened the Spidercrab. Fortunately, the British linguistic superiority asserted itself and the little aircraft's name was changed to the more fearsome Vampire.

John Cunningham disliked his nickname, "Cat's Eyes" Cunningham, given to him by the wartime media for his ability to find enemy aircraft in the dark. A night fighter pilot, he knew that it was because of the secret interception radar that his Beaufighter carried. But to emulate him, thousands of British children ate their carrots. In 1944, he was called to Hatfield to test-fly a new aircraft, the Vampire. It was an absolute revelation, he would later say. Cunningham returned to de Havilland after the war, and in 1947, without a pressure suit, he broke the world's height record by taking a Vampire up to 18,075 metres.

The Gloster Meteor, the first British jet fighter, was operational in time for the Second World War and, by shooting down V1 bombs, received all the glory. But six months after it first flew, on September 20, 1943, de Havilland put the Vampire into the air. Because there was the shortage of metal and no time to experiment with alloys, for the airframe the company used its expertise with balsa and plywood, learned from building the Mosquito. The body was constructed around the jet engine and this permitted an uncomplicated, egg-shaped fuselage mould-

DH Vampire.

ed into a twin boom configuration, influenced no doubt by the P-38 Lightning. The twin booms allowed for a short tail pipe, which, given the weakness of the rudimentary H1 turbojet engine, furnished as much thrust as possible. Subsequent modifications to the basic design were the more powerful Goblin (and sometimes Nene) engines, with successive models having pressurization, a bubble canopy, and underwing fuel tanks.

For its size and simplicity, the Vampire was a historic aircraft: it introduced the modern era to air forces worldwide. It was the transitional aircraft, taking pilots from the piston engine to the jet, from the Wright Brothers to Frank Whittle. Gone were the complicated propeller pitch controls and torque swing on takeoff. If flying without propellers wasn't novel enough, it was also the first pressurized fighter and the first tricycle undercarriage, which many would later have. The naval version, the Sea Vampire, was the first jet to land on an aircraft carrier when Royal Navy Lieutenant Commander E.M. Brown alighted on HMS *Ocean* on December 3, 1945. It was also the first jet aircraft to fly the Atlantic Ocean when on July 14, 1948, six Vampires of RAF 54 squadron landed at Goose Bay, Labrador from England.

As with all early jets, the Vampire's thirst for jet fuel was insatiable and its range was poor. Everything had to be done quickly, especially cockpit checks prior to take off and landing — there was no fuel to waste. The DH 100 Mark III had larger wing fuel tanks, causing the fin and rudder to be reshaped. Armament consisted of four 20-mm cannon in the nose, sometimes unnerving to pilots used to guns in the wings. Later, when it became a ground attack fighter, it carried 21,000 pounds or 8 rockets. The Vampire was bought by fifteen countries, including Australia, New Zealand, Sweden, India, and Switzerland, all eager to replace their piston aircraft with jets.

Courtesy of the DND

DH Vampire.

The post-war RCAF was just as eager and would have liked its jet fighters locally made. But in 1948, such technology in Canada was its infancy; the first Canadian-built jet engine, the Chinook, would only be tested by Avro Canada on March 17 of that year, and the Avro CF100 and the Canadair Sabre were both years away. The RCAF evaluated the Meteor as an interim jet fighter, but decided on the Vampire, taking delivery of the first of eighty-six Mark IIIs on January 23, 1948. All of the RCAF Vampires were given over to auxiliary squadrons except for 421, which was reformed at Chatham, New Brunswick, as the second jet fighter squadron.

This was a pilot's aeroplane, and to the RCAF pilots the Vampire was an extension of their sports cars — it was easy to fly, nimble, and aerobatic. On August 18, 1951, an RCAF aerobatic team flying Vampires performed at the National Air Races at Detroit. After their Mustangs, it was level with the runway and vibrationless, and starting the Goblin was simplicity itself — except for fear of flameouts, common to the early jets. Unlike the Meteor's Derwent engines, the Vampire's Goblin could not be relit and the pilot's choice was either bail out or force land — quickly.

With more Sabres and CF100s coming into squadron service, the Vampires were gradually retired and in 1958 sold off to the Mexican Air Force. Vampires saw action in wars distant from Canada and Britain: they were in the India/Pakistan conflicts and were used to great effect by the Rhodesian Air Force against guerrilla movements after 1972. Fortunately for museums, air shows, and wealthy aircraft collectors, the Swiss Air Force operated their Vampire T.55s until 2000 before selling them off. Now in private hands, they keep the memory of the little jet fighter alive.

AVRO C-102 JETLINER

"There is a tide in the affairs of men, which taken at the flood ..." Shakespeare could have been describing the historic opportunity that the Avro Jetliner presented Canada with, to lead the world in commercial aviation.

At the Second World War's end, British aircraft manufacturer A.V. Roe bought the Victory Aircraft plant at Malton, hoping to produce its Lincoln XV bomber and/or York transport there. When both projects fell through, it kept busy with two advanced fighter designs, the famous Arrow and the CF-100, Canada's first jet fighter. But Avro Canada (as A.V. Roe had become) still hoped to get into commercial aviation with another equally maverick aircraft, a jet-powered airliner. To understand how courageous such a project was one must remember that only two countries had gas turbine engines then: defeated Germany and Britain, where both the Gloster Meteor and the de Havilland Vampire were flying. In Canada, jet engine technology was being studied by Paul Dilworth and Ken Tupper, research scientists at the National Research Council in Ottawa, but it was still in the theoretical stage. In May 1946, A.V. Roe bought Turbo Research Ltd., the only company in Canada in the field.

Avro's design for the C-102 was far ahead of its time. It could take up to fifty passengers, cruise at 450 miles per hour, and even today would not look out of place at an airport. The problem was the selection of available jet engines. Rolls Royce had just developed the AJ-65 Avon, an axial-flow jet engine, and Ernest Hives, the company chairman, gave Avro some good news. The engine was on the British Air Ministry's classified list, but Hives promised that it was about to be declassified and that it could be built in Canada for the proposed C-102 Jetliner.

Avro Jetliner.

The jet transport project appealed to C.D. Howe, the highly influential Minister for Reconstruction. An engineer himself, Howe was also an aviation enthusiast — he had almost single-handedly created Trans Canada Airlines. But above all, he was a patriot, and if Canada could build the world's first jet transport, then he was all for it. Avro knew that there would soon be a market for the Jetliner, as TCA was looking for a replacement for its DC-3s. Such an aircraft had to have a range capability so that a full passenger payload could be carried from Toronto to Winnipeg against headwinds, an exercise within the Jetliner's capacity.

At first all went well. The Jetliner was designed around two AJ65 Avon engines, and as Hives had predicted, they were taken off the Secret list by the British government. In April 1946, approval was given by Ottawa for TCA to purchase the Avro C-102 (without its two AJ65 engines) at a price of $350,000 each. No one knew how much Rolls Royce would charge for the AJ65s, but by now Howe was beginning to think that it would be a lot more than he was willing to pay. Yet rather than pull out of the program, he advanced Avro $1.5 million to continue the development of the Jetliner.

When the British government put the AJ65s back on the Secret list, Avro found itself with an advanced airframe but no engines. The only other powerplant available was the civil version of what drove the Gloster Meteor: the Rolls Royce Derwent. The Derwent had a centrifugal compressor, was less powerful, heavier, and consumed 13 percent more fuel at cruising speed than the AJ65. To get the equivalent power output, the Jetliner had to be redesigned around four Derwents. This increased the aircraft's gross weight from 45,000 to 55,000 pounds, reducing the range so that it could no longer fly between Toronto and Winnipeg with a reasonable payload. All of this, with a critical centre of gravity problem (one check pilot commented that it made a difference on which side you parted your hair), caused Howe and TCA to conclude that the Jetliner could no longer meet requirements, and they withdrew from participation.

Avro tried to put the best face it could on this setback. Four engines, they said, made the Jetliner more reliable, and the Derwents had over one hundred thousand hours of military service already. The Gloster Meteor had captured the world speed record, and compared with the unknown AJ65, the Derwent was the best-tested jet engine in the world. Avro was even able to install the four Derwents under the Jetliner's wing so that the wing spar structure remained unbroken.

But without Howe's endorsement, the Jetliner was doomed. Trying to get out of debt, TCA could not afford the problems associated with a developmental aircraft, and the noisy saga of the Rolls Royce Merlin engines that the airline used in its North Stars did nothing to endear the British manufacturer to them. Desperate for a sale, Avro proposed other specifications and redesigned the Jetliner, offering it once more to TCA but at the higher price of $750,000. The prototype (and only) Jetliner did fly on August 10, 1949 — thirteen days after the Comet — but by then TCA had its North Stars and they were priced at $660,000 each. This did not stop the manufacturer, wrote Gordon McGregor, the airline's new president, from periodically intimating that TCA was buying the Jetliner, in its efforts to make sales elsewhere.

On March 10, 1950, the Jetliner, accompanied by the CF-100, flew from Toronto to Ottawa, landing at Rockliffe for an official reception. For a flight that normally took one hour and forty minutes, the Jetliner had flown in thirty-six minutes. It was a record, and everyone who attended, from the Governor General to the visiting Prince Bernhard of the Netherlands, was impressed. It was a superb aircraft. But none of the dignitaries ran TCA.

Avro then took the Jetliner on a demonstration tour below the border, and without a jet transport of their own, the Americans loved it. National Airlines wanted four, and the United States Air Force, which rarely bought non-American aircraft, allocated funds for a possible purchase of twenty Jetliners for high-altitude navigational training. Trans World Airlines owner Howard Hughes had it flown to his airport at Culver City, California, where he even took the controls himself. To make the aircraft more appealing to the Americans, Avro went to Pratt & Whitney Canada to see if they would manufacture the J42, a licence-built Nene engine. Pratt & Whitney agreed to supply one set of J42 engines but, as it was being upgraded, held out little hope for the project.

(l-r) Walter Deisher, vice-president and general manager of A.V. Roe Canada Ltd., the Right Honourable C.D. Howe, and the U.S. Ambassador to Canada, with the Avro Jetliner. Malton, Ontario, August 1949.

Ultimately, with Ottawa uninterested and TCA not buying, fascination for the Jetliner fizzled out and it was effectively orphaned. In 1951, the federal government ordered that all work on the C-102 cease and that Avro concentrate on building the fighter needed for Canada's Cold War defence, the CF-100. In the end, TCA's forecast of performance proved to be accurate, and all other airlines who examined the aircraft came to the same conclusion. Had the AJ65 engine been used, the Avro C-102 would have been a commercial success. That Ottawa did not keep the project going until the Avon was declassified was to its discredit. Perhaps it had come too late into the life of C.D. Howe, who once said that Canadians could do anything they set their minds to. In years to come, historians would say that Avro was to blame for not judging the market first — that the aviation industry wasn't a "build it and they will come" one. But through the 1960s, when the engineers at Avro watched million of dollars of federal funding poured into aeronautical projects that were dubious at best, they could only wonder what might have been.

On November 23, 1956, after years as an exhibition curiosity and chase plane for the CF-100, the Jetliner flew for the last time. It was cut up on December 13, and today, only its cockpit section remains at the National Aviation Museum in Ottawa. A year after its demise, Boeing had the first production 707 in the air, and Canada had lost out on making aeronautical history. "Fear no more the frown o' the great ..." Shakespeare might have been writing of the Jetliner's fate.

Courtesy of the National Archives of Canada

Avro Jetliner.

B-25 Mitchell.

NORTH AMERICAN B-25 MITCHELL

I t is almost dawn. An engine starts coughing, followed by another, and then many more. Director Mike Nichols is about to introduce his movie audience to the characters Yossarian, Colonel Korn, and General Dreedle — and eighteen drab B-25 Mitchells. Based on Joseph Hellers' book of the same name, *Catch-22* is a black comedy about a group of flyers in Italy in 1944. To get to this point, in 1978 Frank Tallman scoured the United States for B-25s, worrying that because of their age, they would, in best star tradition, be difficult and not perform. He needn't have worried. The B-25 could always be counted on, beginning with that gusty morning of April 22, 1942, when Colonel James Doolittle flew one off the aircraft carrier USS *Hornet*, bound for Tokyo. Later the Mitchell would also play itself in the movie recreation of the raid, *Thirty Seconds Over Tokyo*.

Chief of Staff General Hap Arnold had one. So did General Dwight D. Eisenhower. Howard Hughes raced one for twenty years. Heiress Barbara Hutton gave one to her husband, complete with king-size bed and leather-upholstered toilet seat. After the war, Jimmy Doolittle, now on the board of Gulf Oil, used a B-25 as his executive aircraft. When they were American allies, the Red Air Force was given eight hundred Mitchells. When they were not, the remaining aircraft were code-named Bank by NATO. There were so many surplus that B-25Js were selected by the CIA as its covert aircraft of choice, used to help rebels in Cuba, the Congo, and Haiti. As late as 1962, the aircraft was on the rosters of the air forces of Peru, Chile, Indonesia, Bolivia, Brazil, and Canada.

In comparison with its contemporary, the B-26 Marauder, the B-25 was boxy and crude in design. Though both originated in response to the same U.S. Army Air Corp requirement for a medium bomber, the two aircraft were as different as the B-17 was from the B-24 Liberator. The B-25 was all rivets and slabsheets of aluminum; the B-26 was all curves, right up to its one-piece plastic bombardier's nose. The B-25 was as rugged and stable as a C-47, which was why it was assigned the Pacific theatre of island airfields. A hotrod aircraft, the B-26's stubby wings magnified its engines and propellers, a thoroughbred that was hard to fly and suited to the European theatre. Named for General Billy Mitchell, the 1920s advocate of strategic bombing, the B-25 was the United States' answer to the German Heinkel He 111 and the

B-25 Mitchell.

British Vickers Wellington. It evolved into a gunship with .50-calibre machine guns in the nose for strafing as part of the Air Apaches 345 Bomb Group at the Battle of the Bismarck Sea because the United States had few fighters in the area then. When the war ended the Marauders were thought too difficult to fly and left in Europe to be broken up. The B-25s were grabbed as rich men's transports, fire bombers, and movie stars. There were so many available that at one time every Air National Guard Unit flew them.

The RCAF used some Mitchells during the war, all allocated from RAF stocks. In 1944, equipped with ex-British Mitchell IIs, No. 13 Photographic squadron (later 413 P) was formed at Rockliffe, Ontario, for high-altitude photography. Post-war, needing a light transport and multi-engine trainer, the RCAF accepted seventy-five of the J model from USAF stocks, auxiliary squadrons Nos. 418 and 406 operating them until 1958. Air Transport Command's 12 Squadron also used Mitchell IIIs. While there are three B-25J models on static display at air museums in Canada, the Canadian Warplane Heritage flies the only operational one.

DE HAVILLAND COMET

E ven the Society of British Aircraft Constructors had never seen anything like it. For the last four years they had met at Farnborough, the home of British aviation, to stage an air show. This year, on September 6, 1949, the silver airliner G-ALVG streaked over the airfield — propellerless! It was so silent that there were many who thought that the engines had cut out. From this minute on, every piston-engine airliner in the world was rendered obsolete. Not just a revolutionary design, the de Havilland 106 Comet carried with it the hopes of the British in defeating the Americans in commercial aviation. One of the cheering audience at Farnborough was the president of Canadian Pacific Airlines, Grant McConachie. So fascinated was he with the metallic silver aircraft overhead that he immediately left Farnborough for the de Havilland plant at Hatfield. There, impulsively, he told them that he wanted two Comets, please.

An outcome of the wartime Brabazon Committees' recommendations to compete with the United States, the Comet was the brainchild of many at de Havilland. The British government asked for a high-speed mail plane to compete with the American piston-engine airliners on the Atlantic. But Geoffrey de Havilland realized that his country's only prospect lay in leaping into a wholly new area — gas turbine power. By the war's end, the modest proposal had grown to a thirty-six-passenger, four-jet propulsion airliner. R.E. Bishop, who had designed both the earlier Comet DH 88 racer and the Mosquito, contributed the graceful shape of the fuselage; captured German research was responsible for the slightly swept-back wings. Four Ghost 50 engines were buried in the wing roots and from their centrifugal compressor came the pressurization and de-icing of the airframe. Initially called the Brabazon Type IV, the prototype was named Comet in 1947, the year that BOAC ordered eight.

DH Comet.

Courtesy of the DND

Courtesy of the DND

DH Comet.

So much was done in secrecy to protect the aircraft from the Americans that it was only when John "Cat's Eyes" Cunningham flew the prototype on July 27, 1949, (it was his birthday) for thirty minutes that the world discovered that it had entered the Jet Age. One must remember that until this point only very select test pilots had actually flown jet aircraft. With its Comet, de Havilland proposed to transport thirty-six to forty civilians — from grandmothers to children — at a cruising speed of 490 miles per hour. McConachie wanted the Comet because CPA's rival, the government-owned Trans Canada Airlines, had nothing like it. It would be the perfect draw for CPA's new route from Vancouver to Sydney, Australia. Wasting no time, he talked the board of directors at Canadian Pacific into buying two. They would be based at Sydney and fly between Australia and Hawaii, refuelling at Canton Island and Fiji. He worked his persuasive magic so quickly that on November 28, 1949, there was a scale model of a Comet 1A Empress of Vancouver resplendent in the CPA Canada Goose logo on the boardroom table at the company headquarters at Windsor Station in Montreal. The Canadian Pacific directors, railway men all, could not fail to be impressed. At $1.5 million each, the two CPA Comet 1As, CF-CUN and CF-CUM, were expensive, and his engineers warned him that their range was too limited for the Pacific. But McConachie had never heeded the advice of accountants and engineers in his life and wasn't about to start now.

BOAC naturally got the first production Comet in March 1951 to begin crew training, and so many record-breaking flights were carried out that no one could doubt that all three — Britain, de Havilland, and the Comet — were far ahead of the rest of the world. On May 2, 1952, the first paying passengers were flown from London to Johannesburg, followed by an August 11 flight to Colombo, Ceylon (now Sri Lanka) via Rome, Beirut, Karachi, and Bombay. There was no lack of customers, especially when de Haviland brought out the Avon engine Comets 2 and 3 for transatlantic flights. Air India, Air France, South African Airways, UAT, Japan Air Lines, and (the sweetest victory of all) Pan American all placed orders; another assembly line was opened at the de Havilland plant at Chester. Even the RCAF, not normally in the vanguard of untried aircraft acquisitions, ordered two Comet 1As in December 1951, to be used as high-speed VIP and troop transports.

In January 1953, as the two CPA Comets were nearing completion, the most senior CPA captains were selected for training at Hatfield. McConachie wanted as much publicity as he could from his purchases and ordered that the first Comet, CF-CUN *Empress of Hawaii*, be ferried to Sydney as fast as possible — breaking any speed records it could. The airliner took off from Heathrow on March 1, making for Beirut and then Karachi. In the early morning of March 3, the heavily laden CF-CUN was taking off from Karachi Airport when it hit the culvert of a drainage ditch at the end of the runway and smashed into the embankment. With all that fuel aboard, the explosion could be seen for miles around. A combination of pilot error and the inadequate Ghost engines was blamed. Such mishaps happened to all new aircraft. But the Comet's fatal spiral had just begun. Almost exactly a year after the inaugural flight, on May 2, 1953, BOAC's G-ALYV broke up on takeoff from Calcutta Airport. It was the strong downward gusts they said, and the Comet's popularity remained unaffected.

By now the two RCAF Comet 1As were on schedule for delivery, the first arriving at Uplands, Ottawa, on May 29 and the second in June of 1953. Both were operated by 412 Transport Squadron, then based at Rockliffe. On September 1, 1955, when 412 was moved to Uplands, the five hangars, 10 to 14, were built to accommodate them. The second CPA Comet, CF-CUM, ready in August, was not taken up by the airline and instead went to BOAC as G-ANAV.

In November, McConachie convinced his directors to buy three Comet 2s with the upgraded Avon engines and longer range. On January 24, 1954, BOAC Comet G-ALYP broke up almost over the island of Elba. Modifications were made to all Comets, but on April 8, 1955, when BOAC's G-ALYY crashed into the Mediterranean, the

DH Comet.

Courtesy of the DND

Courtesy of the National Archives of Canada

Grant McConachie tries the controls of the then-novel Boeing 747 during a demonstration flight from Seattle to Tokyo.

jet airliners were grounded. By the time tests at Farnborough showed that metal fatigue caused by a fuselage fracture near a window had ripped the plane apart, the Comet had lost its appeal. No one wanted to fly in one. The engineers and accountants at CPA must have been relieved — the consequences of a Comet breaking up over the trackless Pacific were too horrible to contemplate.

It would be April 27, 1958, before another de Havilland Comet was in the headlines. On October 4, BOAC put its Comet Mark 4 on the Atlantic run, narrowly beating Pan American's first Boeing 707 by twenty-three days. Uneconomical when compared with the 707, and never meant for the Atlantic, the Mark 4 would normally have to refuel at Gander on the run, and even BOAC was quick to use the new Boeing 707s instead. Only seventy-four Comet Mark 4 were built, the last 4Cs becoming testbeds for the maritime patrol aircraft, the RAFs Nimrod. But by now, Grant McConachie had already bought DC-8s, and in 1965, as enthusiastic as he had once been over the Comet, the CPA president put a down payment of $300,000 on three as yet-unbuilt SST airliners.

The RCAF's two 1As, 5301 and 5302, were returned to Hatfield to be rebuilt as 1XBs. This entailed reinforcing the fuselage, upgrading the Ghost engines, and converting from square to oval windows. They re-entered RCAF service in September 1957. While mainly VIP and troop transports, they were also used to test the NORAD defence systems. Finally, on October 3, 1963, the Comet was retired from the RCAF and stored for disposal. Unfortunately, although it was the RCAF's first jet transport, nothing was done to preserve one. Comet 5301 was scrapped in October 1964, and 5302, re-registered as CF-SVR in early 1965, was sold to Buddy Reid of Miami as N373S. After a time in storage there, it too was scrapped.

The only part of a Canadian Pacific Airlines Comet that remains is the cockpit of BOAC G-ANAV, formerly CF-CUM, on display at the Science Museum in London. As to those who had given it birth, Geoffrey de Havilland died on May 26, 1965, R.E. Bishop on June 11, 1989, and John "Cat's Eyes" Cunningham on July 21, 2002. They had provided the world with its first jet airliner.

"Definitely a CAVU day," painting by Neil Aird. Steve Konopelky's DHC-2 Beaver C-FIDF sets out from the dock at Lillabelle Lake, Cochrane, Ontario.

Courtesy of Neil Aird

DE HAVILLAND CANADA BEAVER

"Friends look at me quizzically when I tell them I am going hunting, then scratch their heads when I mention that it is not the furry *Castoridae canadensis* that I pursue, but the venerable, metal-winged DHC-2. I nearly always come home with a log book full of numbers, a few rolls of exposed film; these are my trophies!" Kingston artist Neil Aird, a plane spotter in his youth, has tracked down the whereabouts of almost every Beaver made, immortalizing them on his website, in his photos and paintings, and, to commemorate the aircraft's Golden Anniversary in 1997, in gold.

When test pilot Russ Bannock took the Beaver prototype, CF-FHB-X, into the air on August 16, 1947, he must have had little idea that he was flying a national treasure. For if the French gave the world haute cuisine, the Italians opera, and the British the language of Shakespeare and Milton, Canada's contribution to civilization is undoubtedly the de Havilland Beaver. It ranks with other national icons such as centre ice at the old Montreal Forum, poutine, the moose, or the Musical Ride of the Mounties.

The Beaver's fathers were Phil Garratt, de Havilland Canada's managing director; W. Jaimiuk, of the PZL fighter fame; W.D. Hunter, who managed Canadian production of Mosquitos; and Fred Buller, who ran the design office. (That the registration of the prototype now owned by the Canada Aviation Museum was CF-FHB-X is no accident. It was the initials of Fred Buller.) At the end of the Second World War, de Havilland Canada cast about for a niche market to enter. Its parent in Britain was preoccupied with building the Comet, and the Canadian branch couldn't compete with Boeing, Douglas, or even Avro or Canadair. Its first post-war effort, the DHC-1 Chipmunk, was creditable but hardly earth-shattering. It flew on May 22, 1946, but competing as it was with thousands of war surplus trainers, sales were initially slow.

Garratt heard that the Ontario Provincial Air Service and various small bush airlines were looking to replace their Norsemen and flying boats and that the bush plane market offered some possibilities. He took a gamble, and the company's first design for a bush plane was a stretched version of the Chipmunk powered by the 145-horsepower in-line de Havilland Gipsy Major 8 engine. De Havilland polled visiting bush pilots about what they wanted in an aircraft,

C-GAQX at Port MacNeill, British Columbia, May 1989.

and the results showed that it definitely wasn't speed or streamlined looks. What was needed in the bush was an all-aluminum, easily accessible aircraft with short takeoff and landing characteristics and plentiful energy and lift.

It was obvious that the Gipsy engine wasn't powerful enough to do all of this, and what saved de Havilland was the news that Pratt & Whitney Canada had located a supply of war surplus P & W Wasp Juniors that had never been used. The radial-engine Wasp was more powerful than the Gipsy, and to get the lift, a thick wing and new flap system was used. The fuselage was made entirely of aluminum, which would be much less trouble in the bush than the fabric and wood of older aircraft. It was to have four doors: two for crew and two larger ones for cargo.

When he test-flew the aircraft, George Neal discovered all kinds of advantages. With a full load and no wind, it could take off in a run of 650 feet within 10 seconds; fully loaded, it could clear a 50-foot obstacle in 1,000 feet; the floatplane version needed a 15-second run to take off. The DHC-2 Beaver was the first of the modern bush planes, and it set the bar high for contemporaries like the Fairchild Husky. Pilots who had learned to fly Fokker Universals loved the convenience of the new aircraft. Besides the metal body and the four doors, the elevator trim control worked logically — backward for nose up and forward for nose down. They no longer needed a stepladder for refuelling. The wide-track undercarriage smoothed out difficulties on rough strips. The control was well balanced. It trimmed easily for hands-off flight. It refused to stall. It was easy to land at 70 miles per hour with full flap. It climbed fastest from a three-point attitude, ascending almost helicopter-like.

The Ontario Provincial Air Service bought sixteen Beavers, and other provincial governments followed. The early models cost $25,000 and were snapped up by mining companies and regional airlines alike. The Australians acquired five for the Australian National Antarctic Research Expedition between 1955 and 1961. Three were destroyed in blizzards by the time the Antarctic Flight was disbanded in 1964. Russ Bannock demonstrated it to the United States military, where as the RU-6A it became the preferred mode of transport for officers, earning the name "the General's Jeep." In total, the U.S. Army and Air Force would buy 970 Beavers, 200 of which were designated L-20As to be flown in aerial evacuation of litter and ambulatory patients. Other missions that the Beaver was used in included courier service, passenger transport, light cargo hauling, reconnaissance, rescue, and aerial photography. In Vietnam, the U.S. Army used some Beavers for locating enemy wireless transmitters, distinguished by the long vertical aerials mounted on the leading edges of each outer wing, a dangerous process whereby the slow moving aircraft had to fly directly over the enemy to pinpoint the transmitter.

By the time the last Beaver was made at Downsview in 1967, a total of 1,692 had been built and sold to 63 countries. In 1987, the Canadian Engineering Society, to mark its centenary, proclaimed the Beaver as one of Ten Outstanding

Courtesy of Rich Hulina

C-GFDS of Knobby's of Sioux Lookout, Ontario.

Engineering Achievements. It was one of two manufactured products selected (the other was the snowmobile). When the Royal Canadian Mint began a ten coin series in 1995 featuring Powered Flight in Canada, coin three was the Silver Dart with J.A.D. McCurdy and F.W. (Casey) Baldwin, and coin four was the de Havilland Beaver with Phillip Garratt.

Aird, who designed the Twin Otter and Dash Eight coins for the Mint, has, in his own words, an extreme fondness for bush planes, aircraft that fly in the world's wild places, particularly in Canada's North. The truth, he admits, is that passionless aircraft spotting is nothing but an oxymoron. "Out of this passion I developed a love for the Beaver; I acknowledge it as a 'passion' or 'love affair'. I know it's not infatuation, in fact I'm sure it's the real thing! I consider the Beaver as Canadian as the ubiquitous red or green canoe on a lake, albeit creating a little more noise when in motion!" He continues:

> The sound of the droning radial engine and large propeller of the Beaver is music to my ears. I had always wanted to do extensive research on a specific aircraft type. When I knew that the Beaver was going to turn 50 in 1997, I thought I would concentrate my efforts on this historic machine. I chose this aircraft for several reasons. Both the Beaver and myself were of about the same vintage. I wanted an aircraft that was still prevalent in my neck of the woods, not some rare type found only in Europe or Africa (although Beavers are found on both those continents). Since many are still active in Canada and the United States in particular, the Beaver was the obvious choice.

Most Beavers, after they were built in Toronto, migrated to many countries — about 63 nations, in fact. In later life they seem to return to their place of birth, or at least the continent where they were built. They come back, not to retire, but to get a new lease on life, soon to be busy as a beaver, hard at work again. Beavers occasionally return under their own power, more often resting in containers on ships from distant places like Australia, New Zealand, and recently, Bangladesh. They come home in good shape, in bad shape, in large and small pieces, often very bent! Several companies in Western Canada, the Pacific Northwest of the United States (Washington State in particular), and Minnesota specialize in rebuilding the stalwart Beaver to new, or better than new, condition.

When I mention the Beaver in conversation, I'm always amazed how many people have a Beaver story, usually about a time when they flew in one, or about someone they knew who had worked with them. We owe a lot to this fine machine: it helped open Canada's more remote areas, and it has saved many lives in its air ambulance role. It has been used by medical personnel, surveyors, miners, and prospectors.

There are many who would agree with Aird — and not all of them in Canada. Like the maple leaf and the cry of the loon, the de Havilland Beaver is Canada.

LOCKHEED NEPTUNE

The Lockheed Neptune had the misfortune to arrive too late for the Second World War and too early for the Soviet submarine threat. Worse, it served in between two better known aircraft — the Hudson and the Orion. In RCAF service, it was a stopgap measure until the Canadair Arguses became operational.

By the end of the Second World War, Lockheed was riding high on its Hudson maritime patrol aircraft, a militarized version of its Electra airliner. By stretching it and adding more powerful engines it capitalized on the basic design and produced a pair of maritime aircraft — the Ventura and the Harpoon. In 1943, Lockheed sought to take this to the ultimate with a longer range, better armed, and radar-equipped patrol aircraft. The U.S. Navy approved and awarded the company the contract.

The first Neptune flew on May 17, 1945 — bulbous nosed, twin Wright Cyclone R-3359-8 piston engines, and, unlike the Harpoon, with a single large vertical tailplane and tricycle landing gear. Six machine guns in pairs in the nose, dorsal turret, and tail provided the armament. Most prominent was the radome between the nosewheel and bomb bay with the search radar. Designed for longer patrols than its predecessors, it was comfortably equipped (for 1945); the wings, which could provide some floatation if the Neptune was ditched, were fitted with de-icers.

At first for use as maritime patrol/anti-submarine aircraft the RCAF brought its Lancasters out of storage, before buying twenty-five of the last model Neptune in 1955. This was the P2V-7 with updated equipment that required more power for takeoff. Its Turbo Compound Wright Cyclone engines with methanol-water injection were supplemented

Lockheed Neptune.

Courtesy of the DND

Lockheed Neptune.

with Westinghouse J-34 turbojet engines — one of the few aircraft to have that combination. The P2V-7s were also bought by Australia, France, and the Netherlands and made by Kawasaki in Japan.

It had an enlarged canopy, longer fuselage, clear nose, and MAD boom band wingtip tanks. Like Henry Ford's cars, it came in only one colour: midnight blue. Repainted later in the RCAF's maritime colour scheme of grey and white, Neptunes served in Maritime Air Command from 1955 until 1968. Initially, 404 and 405 Squadron operating out of Greenwood, Nova Scotia, and later Summerside, Prince Edward Island, flew them, closely followed by 407 Squadron at Comox, British Columbia. Although anti-submarine warfare was the Neptune's main role, the RCAF armed some with unguided rockets for anti-shipping. When the Canadair Arguses became available, the last Neptunes were sent to 407 Squadron in Comox until they received the new aircraft.

Fairchild C-119.

Courtesy of the DND

FAIRCHILD C-119

It was noisy and, with the clamshell doors open, drafty, but the RCAF never had an accident with one. Called the Flying Boxcar, the Packet, and the Dollar 19, the Fairchild aircraft was a twin-engine, twin-boom freight and troop carrier. Developed from the company's wartime C-82 utilitarian freighter, the C-119 was refined in1947 into the familiar shape. Too late for the Second World War and just missing out on the Berlin Airlift, the Packet served the United States military well in the Korean War and in the Space Race — and in extreme old age as a gunship in Vietnam.

With a ten-thousand-pound payload, the C-119 could carry sixty-two paratroops or up to thirty-five stretcher cases. The rear opening clamshell doors were designed for quick, efficient handling, and it was sold to India, Belgium, Italy, and Canada. With its Dakotas and North Stars inadequate for long-range heavy transport operations, the RCAF bought thirty-five C-119s, equipping 435 and 436 Squadrons in Air Transport Command; the first arrived in September 1952. This would allow Canada to provide the United Nations with air transport, proven in 1956 when 435 based at Edmonton flew its C-119s to Capodochino, Italy, to support UNEF peace keeping operations in Egypt. Both squadrons were also tasked with airlifting cargo to the joint American-Canadian Arctic outposts through the early Cold War. As these resupply flights could only be made in the spring and fall, the C-119s would land on sea ice or primitive land runways. Much of the resupply was done by parachute drop, an exercise the C-119 excelled in at the paratroop school at CJATC Rivers.

In the United States, one C-119 was modified for mid-air retrieval of space capsules, and on August 19,1960, one made the first mid-air recovery of a capsule entering the earth's atmosphere by snagging the parachute of the

Discoverer satellite south of Hawaii. By the time the war in Vietnam had begun, the C-119 was overshadowed by the C-130 and downgraded to being expendable. C-47s called Puff the Magic Dragons and AC-119s called Shadows carried out nightly reconnaissance missions as illuminators and gunships. The Vietnamese moved their supplies by truck at night, and both aircraft would drop flares or use their xenon lights over a suspected target then fire miniguns and later cannon into the illuminated area.

Fairchild made 1,112 C-119s in all versions. Its death knell in the RCAF sounded on October 28, 1960, when the first C-130 Hercules were delivered to 435

Courtesy of the DND

Fairchild C-119.

Squadron. By 1964, all C-119s had been sent to storage at the former RCAF Station at Saskatoon, where Bristol Aerospace had the contract to dispose of them. Eventually there were thirty Flying Boxcars sitting on the tarmac by Hangars 4 and 5 until all were finally sold in 1967. But the new owners could not obtain commercial certification for their use, and since the terms of sale included the removal from the Bristol site, the C-119s were towed to a distant side of the airport until gradually they disappeared to parts unknown, the last in 1974.

McDONNELL BANSHEE

In Gaelic folklore, a Banshee is the female spirit whose wail warns of imminent death. In Canadian tradition the Banshee's wail warned of the death of carrier-borne fighters in the Royal Canadian Navy, and of the aircraft carriers themselves. It was the name that McDonnell, following its tradition of naming its aircraft after ephemeral objects, gave its F2H-1, one of the U.S. Navy's earliest carrier-borne jet fighters. When U.S. Navy Banshees served in Korea as bomber escorts, they were found to be rugged, reliable, and, for a shipboard fighter, to have excellent range. Encouraged by this, McDonnell set about designing the F-4 Phantom.

In 1951, the Royal Canadian Navy was looking to replace its Hawker Sea Furies with enough jet fighters to equip two squadrons, and favoured the Banshee. The federal government was still unsure about its commitment towards naval aviation, and it suited Ottawa to have the loan of two ex-British aircraft carriers, redesignated HMCS *Warrior* and HMCS *Magnificent*, for the RCN. Now the British pushed Ottawa to buy from them the surplus Royal Navy carrier HMS *Powerful*, completed in 1946, from which jet fighters (hopefully also bought from the British) could be launched. But when that carrier was bought in 1952 and renamed HMCS *Bonaventure*, there was a poor selection of British-built carrier fighters available, giving the Banshee supporters in the RCN an opportunity to press their claim.

This was where the politicians lost their nerve. A second large capital purchase for a branch of the Navy that they saw as marginal in purpose led to lengthy political wrangling. But by the time Ottawa decided on the fighter purchase and allocated the funds, it was too late, as in September 1953 the McDonnell production line for the Banshees shut down. The Canadians then went cap in hand to the U.S. Navy for their used F2H-3 Banshees, and

McDonnell Banshee.

Courtesy of the DND

McDonnell Banshee.

thirty-nine were acquired for two squadrons, VF870 and VF871. To launch them, the *Bonaventure* was fitted with an angled deck and steam catapult.

Nicknamed the Banjo, the thirty-nine aircraft served with the RCN as all-weather interceptors and fighter-bombers from November 26, 1955, to September 12, 1962. Why thirty-nine? It was decided that there would be sixteen (eight per squadron) with another sixteen in reserve and seven for attrition. The fighter had a maximum speed of 580miles per hour at sea level, an initial rate of climb of 6,000 feet per second, and a range of 1,170 miles.

It was armed with four 20-mm Hispano cannon, six Aero 14B 500-pound bombs, and two heat-seeking Sidewinder AIM-9B air-to-air guided missiles. At this time the Philco-made Sidewinder was very secret, and neither the RCAF nor the British had it. Nine feet long, five inches in diameter and, weighing 156 pounds, it lived up to its name, emitting a loud hissing sound as its homing device locked onto a target. Sidewinders would be used to great effect by the Taiwanese against the Communist Chinese and by the Royal Navy's Harriers in the 1982 Falklands War against Argentinian aircraft. It would not be until the CF-18 that a Canadian aircraft would again be equipped with Sidewinders.

The two Banshee squadrons, VF870 and VF871, were amalgamated in 1959 and then paid off in September 1962 as all the Banshees were retired from the RCN. Relegated to scrap by Crown Assets Disposal the following year, they were cut up — just as the Cuban missile crisis took place. Then there were panicked messages from Ottawa to get them flying again. But it was too late: the welding torches had already done their work. The last Banshee was flown to Calgary for use at the Southern Alberta Institute of Technology, and the carrier-borne fighter tradition in the Royal Canadian Navy ended with it.

S-51 Dragonfly

Courtesy of DND

SIKORSKY DRAGONFLY

The H-5 Dragonfly or S-51 always calls to mind the movie *The Bridges at Toko-Ri*, based on the James Michener novel about a U.S. Navy aircraft carrier off the North Korean coast in 1953. While the Grumman Panthers (and Grace Kelly) were the stars of the movie, audiences were also enthralled by the H-5 rescue helicopter (with Mickey Rooney and Earl Holliman as the crew), which successfully plucked downed pilots out of the water and enemy hands. If jet fighters were unusual, then helicopters were more so. Soon after the Second World War the public had heard and read so much about these devices that it was thought that everyone would have one in their driveway.

When Sikorsky built the first R-5 in August 1943, helicopters were still called rotorcraft, hence R-5. It was a step up from the little two-seater R-4, his first helicopter, but still impractical for rescue or transport. In 1946, the powerplant was changed to a 450-horsepower Pratt & Whitney R-985 Wasp Jr., the cabin was enlarged to seat four, and a strong tricycle landing gear was added. It was the first helicopter to be fitted with a rescue hoist and hydraulic pitch control and to carry the pilot and three passengers. The H-5 flew on February 16, 1946, and was the first Sikorsky model to be commercially approved. United Air Lines and Los Angles Airways both bought civil versions called S-51s. There were 214 built altogether in the United States, and in Britain Westland acquired the licence to build 139 from 1948 to 1953, calling them Dragonflies. The British version used a 550-horsepower Alvis Leonides engine instead of the Pratt & Whitney.

The H-5 achieved its greatest fame with the U.S. forces in Korea, picking up downed pilots behind enemy lines and airlifting the wounded from the front lines directly to a MASH (Mobile Army Surgical Hospital) unit. Canadians had flown R-4s with the U.S. Coast Guard in 1945, and the RCAF saw the S-51's potential for SAR. In 1947, it took the plunge, and on April 5, seven S-51s were bought from the Sikorsky plant at Bridgeport, Connecticut, becoming the RCAF's first helicopter. As they had been since 1932 when they sold a Sikosrky flying boat to Canadian Airways, Pratt & Whitney Canada were the Sikorsky agent, and all RCAF S-51s were maintained at Longueuil.

Unlike Bell, Sikorsky had no success selling its S-51s to commercial operators, especially in the bush. Their limited range, experimental nature,

S-51 Dragonfly.

and expense — almost twice the price of other helicopters and conventional bushplanes — made their costs prohibitive. All RCAF S-51s were phased out by 1965, and the first one, RCAF 9601, was given to the National Aeronautical Collection in Ottawa.

CANADAIR COSMO

"The Cosmo/Convair was a very good old truck; very powerful but noisy and very stiff on the controls," remembers Andre Gignac, who flew them in the military. "The Allison engines were very reliable — a tremendous improvement over the original Napier Eland installation. When I first got to the squadron, the old captains sometimes referred to this previous Convair-Napier arrangement as the world's largest single-engine aircraft, a testimony to frequent engine failures in flight. Not so with the T56, though."

When American Airlines wanted a replacement for its DC-3s in 1945, Convair designed the Model 110, a 30–40 passenger aircraft with Pratt & Whitney 2,100 Double Wasps. The airline thought that it could be stretched and Convair obliged with the Model 240, which was also bought by Swissair, KLM, and Northeast Airlines. In 1952, Canadian Pacific Airlines bought five Convairs for its provincial routes, using them as feeders into the Vancouver hub and selling them in Japan in 1964.

But Canada wasn't finished with the Convair yet. Both Canadair and Convair were then owned by General Dynamics, and when the 440 model was phased out of production in the United States in 1957, the jigs and all uncompleted airframes were sold to the Cartierville company. The idea was to build Convairs with British Napier Eland tubroprop engines for Trans Canada Airlines and the RCAF. The deal suited everyone: the work would prop up Canadair, a key employer in the politically important Montreal region, and it would help keep General Dynamics from withdrawing its investment. Napier was pleased at getting FAA certification, which would open the door to commercial sales in the American market, where the turboprop Fokker F-27s were very successful.

Canadair Cosmo.

Courtesy of the DND

Heavy-lift and eggshells: CC-130 and Cosmo.

Except that Gordon McGregor, the president of TCA, was canny enough to know a lemon when he saw one. As he had previously with the Canadair Yukon, McGregor refused to cooperate with Ottawa's plan. Besides, he had already ordered turboprop Viscounts from Vickers, and the RCAF was close to doing the same. The federal government was not able to influence McGregor, but it could exert enough pressure on the RCAF to buy ten turbo-prop Convairs for 412 squadron. It had begun the war as #12 Communications Flight in September 1939, and through the post-war years 412 was tasked with providing air transport for VIPs.

It initially flew Yukons until they were retired in 1968, and then operated a fleet of Falcon executive jets and six Cosmopolitan turbo-prop transports from Hangar #13, CFB Uplands.

Designated the Canadair CL-66 and called the Cosmopolitan or Cosmo, the first built were from the Convair airframes. The first Cosmo for the RCAF (designated CC-109) flew on January 7, 1960. Canadair tried hard to market the aircraft, and there were demonstration tours and talk of building enough to sell to South American airlines. But all of this came to naught, and by 1961, Canadair was hastily distancing itself from the whole Convair project. By now the Napier Eland had proved itself so unreliable and in today's language high maintenance that in 1966, the eight remaining RCAF Cosmos were re-engined with Allison 501-D36 turboprops. Gignac continues:

Although an Ottawa squadron, 412 kept two detachments manned when I was with them. One was in Colorado Springs to serve the transportation needs of NORAD headquarters staff. The permanent radio call sign of that aircraft was Smokey Zero-Two. The aircraft was 109154.

The most outstanding duty, however, was with the Lahr (Germany) detachment. The flying in Europe was mostly Specials (non-scheduled) with only one sked to London-Gatwick once a week (with a stop in Belgium). So we were all over Europe, Africa and the middle-East. We had ground troops in Cyprus, in the Suez canal area and in the Golan Heights so we had lots of flights into that general area. Things would sometimes get a little bit crazy because some Arab countries would not accept a flight plan or provide a clearance into Israeli airspace; or they would refuse to acknowledge that Ben-Gurion airport, in Tel-Aviv, actually existed. It sometimes made for creative flight planning.

The VIP flights from Uplands were called eggshell flights for good reason. The passengers could be the royalty, either visiting like the Queen Mother or local like the Governor General. Every time one of these aircraft would come home, a team of groomers would attack it with rotary buffers till it shone like a mirror. Eventually this started wearing the Alclad protection right off the metal. This bare metal lower portion of the fuselage was eventually painted grey sometime in the '80s, I think — after my time.

Air Canada Captain J. Allan Snowie joined 412 Squadron in October 1970 and flew VIPS to various destinations; the entries in his Cosmo logbook are a Who's Who of the government in microcosm. As part of the eggshell runs, he took African presidents to Ottawa, Prime Minister Pierre Trudeau and wife to Rome, as well as assorted ministers, members of Parliament, Canadian ambassadors, and generals — one of the last specifically to Lourdes, France. Other passenger loads in the Cosmo were lobsters from PEI, band

Leader of the pack: Cosmo and Boeing 707 at CFB Lahr, Germany.

instruments, DND school teachers, and army cadets on sightseeing flights. Some of Snowie's logbook entries for the Cosmo warrant explanation. On one occasion he flew twenty-three drunk sapsuckers from CFB Bagotville to Ottawa. He explains:

The fighter boys in Bagtown ran a rights of spring all-comers each year. Something to do with the sap rising (maple, that is) and intercept operations (on the young ladies of the Saguenay, I believe). The Americans loved it and would appear in droves with various squadrons. Obviously that Cosmo flight was a medevac to return some of the NDHQ lads back to flying their LMD's — large mahogany desks.

He also recalls:

The Convair 580 has been ignored yet it continues today with the superb rebuilds being done by Flightcraft (I believe) in Kelowna. Power assisted controls and glass cockpits ... It was the only aircraft that I've ever flown that would consistently give you smooth landings once you got the touch. Worked very well in the VIP unit.

There are the stories about the engine failures with the original power units. I believe that a certain air vice marshal was aboard during an approach into Downsview when the aircraft suffered a double engine failure. The re-engining to the upside-down Herk units followed immediately — or so the story goes.

Another Cosmo vignette — Two were taken out of service and stored in North Bay. Then sold for lakeside cottages. The new owners apparently made a small fortune when the chagrined CAF had to buy back the wings for the spars.

Vertol H-21

VERTOL H-21

While Sikorsky, Bell, Westland, and Augusta are well known in the world of helicopters, Piasecki isn't. Yet Frank Piasecki was the second American, after Igor Sikorsky, to build and fly a helicopter. In 1943, he flew his own design, the Piasecki-Venzie or PV-2, around Washington to demonstrate it before the military. But although the United States was at war, no one saw its potential, and it was only with the help of Congressman Harry S Truman that the Coast Guard agreed to buy one from him.

The U.S. Navy became interested in helicopters for antisubmarine purposes, and in 1946, the Piasecki Helicopter Corporation built the HRP-2, a large tandem rotor model that was immediately dubbed the Flying Banana. With a strengthened fuselage and 1,425-horsepower Wright Cyclone radial engine as a powerplant it was able to carry two crew and twenty passengers. The world's first tandem rotor transport made its first flight on April 11, 1952, and was an immediate success. Officially called the H-21 Workhorse by the U.S. Army, it was bought by Germany, Japan, Sweden, France (where it was called *banane volante*), and Canada. It was first used in combat by the French Army in Algeria and later the U.S. Army in Vietnam.

Canada bought six H-21As and nine H-21Bs for service with the RCAF in 1955 for communication, SAR, and resupply of the Mid-Canada radar line. Designated the CH-125, all were struck off by January 12, 1972. The Piasecki company underwent a name change in 1955 when the Vertol Aircraft company bought Frank Piasecki out. Vertol continued with the development of tandem dual rotor helicopters, and in 1960 Boeing took it over, making it its helicopter division.

Vertol H-21

VICKERS VANGUARD

Sturdy, structurally sound, with the most spacious cockpit ever in a commercial aircraft, the Vanguard was a pilot's plane. Passengers in the Swinging Sixties thought different: that it was dated and lived in the shadow of the DC-8 and Boeing 707. The pilots knew better, and as one said, the Vanguard was the last of the four-masted square-riggers in a world that had become increasingly modern — and boring.

The Vanguard was born in 1953 when British European Airways looked for a successor to the Viscount. It wanted another turboprop airliner from Vickers that would be faster and bigger but operate at lower costs, in effect a super Viscount. Trans Canada Airlines, similarly impressed with its Viscount 724s, also wanted a larger version for its cross-Canada and high-density routes like Montreal-Toronto and Toronto–New York. Vickers set to work, custom-building an aircraft for both airlines and in effect making them its only customers. BEA favoured a high-wing design like its Elizabethan; foreseeing de-icing problems, TCA did not. Both agreed on a huge cargo hold so that the aircraft could be used on long, continental routes. As a result Vickers drew up the V.900, a low-wing aircraft with a double-bubble fuselage powered by four Rolls Royce Tyne engines. BEA was pleased with the V.900 and, on July 26, 1956, placed an order for twenty.

Ironically, for an aircraft that was to be the last prop airliner, Vickers named it Vanguard, perhaps after HMS *Vanguard*, the British battleship that was to be the last of its kind as well. In keeping with the naval theme, BEA christened each of theirs after other famous Royal Navy ships like Amethyst, Swiftsure, and Dauntless. Vickers hoped that the TCA order would be the breakthrough it needed to get into the American market, where it faced stiff competition

Vickers Vanguard.

Courtesy of Ken Leigh

from Lockheed with its turboprop Electra. As a result, when TCA wanted a greater payload it revised the V.900 design into the larger V.952. They needn't have bothered, for the president of TCA, Gordon McGregor, was determined to have an all-Rolls-Royce-engine fleet, and with the Vanguard he would now have one. On January 31, 1957, he signed an order for 20 Vanguards at a total cost of $67.1 million, with options for more. Although there were still problems with the Tyne engines, on January 20, 1959, Vickers test pilots Jock Bryce and Brian Trubshaw lifted the first Vanguard, G-AOYW, off the runway at Weybridge. While this aircraft was sent to the Sudan for hot trials, the second Vanguard, G-APEA, was flown to Montreal in June 1959, for testing by TCA. Both BEA and TCA were so pleased with it that on May 21, 1960, the first V.952 TCA aircraft, CF-TKA, was airborne and displayed at the Farnborough Air Show.

Then the Tyne engine flaws reasserted themselves — there had been a compressor failure on a BEA London-Athens flight, and all Vanguards were grounded. The TCA engineers were now reluctant to accept the aircraft, but Rolls Royce speedily replaced the compressor discs, and the Vanguard received its Certificate of Airworthiness on December 2, 1960. By December 17, BEA had its first Vanguard on scheduled service, and after extensive testing, on February 1, 1961, TCA inaugurated its first Vanguard service with CF-TKD on the Montreal-Vancouver route via Toronto, Winnipeg, Regina, and Calgary.

Vickers tried hard to find other buyers for its Vanguards, sending it on demonstration tours to the United States, Mexico, South America, the Middle East, and Africa, hoping that former Viscount customers would upgrade to it. But prospective buyers realized that the day of the turboprop had ended, and although as BEA was at pains to explain that the Vanguard was almost as fast as the jet airliner Caravelle, it was jets that passengers wanted. Soon the Comet 4B and Tridents replaced the Vanguards on BEA's high-density routes, and by 1967, more and more, the Vanguards were relegated to the package holiday flights. As a result, when the final Vanguard, CF-TKW, was delivered to TCA/Air Canada on April 3, 1964, the line at Wisley was closed. In total, besides the prototype, forty-three Vanguards were built: twenty for BEA and twenty-three for TCA.

It suffered two crashes, one attributed to pilot error and the other to poor maintenance. On October 27, 1965, a BEA Vanguard attempting to land at Heathrow in the fog overshot the runway and crashed, killing all thirty-six people on board. Six years later on October 2, 1971, another BEA Vanguard broke up in midair over Arselle, Belguim, killing all sixty-three on board. The cause was later found to be failure of the rear pressure bulkhead, which had suffered corrosion from a leaking toilet.

There were no Vanguard crashes in Canada, but although they worked hard, by 1966 the aircraft were perceived as old-fashioned by the public, and in December of that year Air Canada withdrew CF-TKK from passenger operations. Propellers were out in the eyes of the public, no matter if they were attached to a Rolls Royce turbo engine. From January 1969, Air Canada began taking the other Vanguards out of service, trading fifteen in to Lockheed as part exchange for its Tristars. Lockheed resold them to British disposal company Air Holdings, and they ended their days with Scandinavian and Icelandic companies Air Viking, Air Trader, and Thor Air Cargo, before being broken up at Perpignan, France in 1980. Other Air Canada Vanguards were sold directly to the French air company Europe Aero Service. The final scheduled passenger Vanguard flight took place on October 31, 1971, in the Maritimes. BEA converted its Vanguards to be dedicated all-cargo carriers, calling them Merchantmen, and Air Canada did the same with CF-TKK, calling it the Cargoliner, finally retiring it in May 1972. For many pilots, it will always be the last of the square-riggers.

CANADAIR DYNAVERT

The dream of building a craft able to fly vertically as well as horizontally is older than powered flight. The use of Montgolfier balloons in the 1780s led to steerable dirigibles a hundred years later that were propelled first by pedal power and then by lightweight combustion engines. Juan de la Cierva's 1920s autogyro, although it could not take off vertically or hover, was the next attempt toward combining flight on two axes. In the 1950s, the British-built Vertical/Short Takeoff and Landing (V/STOL) Fairey Rotodyne seemed to hold the promise of flight between city centres. It could be landed vertically like a helicopter or on short city centre airfields like an autogyro.

Trans Canada Airlines became interested in the Rotodyne for use on its Vancouver-Victoria route, but its very high initial cost ($2 million per aircraft), coupled with low cruising speed, wasteful fuel consumption, and the noise it made lifting off from those city centres, could not justify any profitability. When British European Airways lost interest for the same reasons, Whitehall cancelled the Fairey Rotodyne in 1962. But the military applications of such an aircraft interested the North Atlantic Treaty Organization, which issued a specification for a V/STOL aircraft for combat, transport, and SAR operations. Convair was testing its Pogo fighter, as was the Bell Helicopter company a tilt-rotor craft, the XV-3. The inherent risk of developing convertiplanes required heavy long-term sponsorship, and the cost overruns would force Convair to stop and Bell to combine forces with Boeing. However this did not seem to deter Canadair, and in 1957 its engineers began researching the design of a tilt-wing V/STOL aircraft. It was an ambitious venture for the Cartierville company, particularly as until then, with the exception of the CL-41 trainer, it had only built aircraft under licence. The company chose to go the tilt-wing way, building a conventional-style air-

Courtesy of Canadair

Canadair Dynavert.

craft body with two engines on a massive wing that the pilot could move horizontally as well as vertically, allowing it to fly like an aircraft or a helicopter.

Canadair sold the CL-84 Dynavert to the federal government on its military application, and in 1963 the National Research Council, the Department of Defence, and the Defence Resource Board agreed to finance and test the unconventional design. The whole field was so new that the company embarked on extensive wind tunnel tests at its plant, even creating a rudimentary flight simulator to keep the cockpit as traditional as possible. The Dynavert's V/STOL powers depended on two 1,500-horsepower Lycoming turboshaft engines driving the massive 14-foot-diameter rotors, which could tilt upward through 100 degrees. If this was not unusual enough, there was a pair of horizontally-mounted contra-rotating propellers mounted on the tail for pitch control, which were stopped when in conventional flight.

The CL-84 made its first VTOL flight on May 7, 1965, followed by a STOL flight in December. The first completely transitional flight took place on the gusty winter day of January 16, 1966, taking about twelve seconds to transit from hover to conventional mode. Tests through the summer showed the CL-84 to be responsive, stable, and capable of lifting almost 3,000 kilograms of fuel and payload in STOL mode and 1,800 kilograms in VTOL. In response to the NATO bid, various military uses were tried, machine guns were attached and fired, there was a SAR exercise, and it was marketed as a gunship and antisubmarine patrol craft.

The prototype crashed September 12, 1967, on a test flight, and three more versions called CL-84-1s were built between 1968 and 1970. These were bigger, more powerful, and could seat twelve. They first flew on February 19, 1970. Canadair worked hard at interesting the Canadian, British, and American military, giving the Dynavert over in 1972 to U.S. naval pilots to fly at the U.S. Naval Air Test Center in Patuxent River, Maryland. But the British were then involved with developing their Harrier, and although one CL-84-1 would be exhaustively demonstrated at the Pentagon building itself and from the aircraft carrier the USS *Guam*, Canadair failed to interest anyone in Washington, London, or Ottawa in it. The tilt-rotor aircraft died at birth. The helicopter lobby in both countries was thought to be too strong. The third Dynavert never reached completion, and the program was cancelled in 1974. One of the CL-84-1s was donated to the Canada Aviation Museum in Ottawa, another to the Western Canada Aviation Museum in Winnipeg.

A commercial failure, the CL-84 proved to be a pioneer in the tilt-rotor concept. After forty years of development and a $5-billion investment, the American government did give official approval to the Bell Boeing Tilt-Rotor concept, and in 1995, production was begun on the V-22 Osprey as a VTOL troop transport for the U.S. Marines. In the commercial field, FAA certification is planned in 2007 for the Bell/Augusta BA 609, which transitions from helicopter to airplane mode in twenty seconds. Its design and the use of Pratt & Whitney Canada PT6C-67A turboshafts gives some connection to the Dynavert.

DHC Caribou.

DE HAVILLAND CANADA CARIBOU

The de Havilland Caribou's special alchemy made it equally at home on a Vietnamese jungle strip and in the tundra. Although not used by the Canadian military in Vietnam during that conflict, it inadvertently became Canada's main contribution to it, serving there with American, South Vietnamese, and Australian forces ... and with the North Vietnamese victors when it ended.

When George Neal and David Fairbanks flew the prototype Caribou at Downsview on July 30, 1958, they could expect no headlines in the media that day. It was the summer of the Avro CF-105 Arrow, then exceeding Mach 1 in level flight, and all attention was on what was taking place at Malton. But for de Havilland Canada, finding markets for the high, inverted gull wing aircraft was a venture just as precarious. With its Beavers and Otters the company had built itself a niche in the short takeoff and landing market, but an aircraft so large and with such limited appeal as the Caribou was almost beyond its scope. There did not seem to be a military market for a mid-size tactical aircraft, and the initial lack of orders almost bankrupted the Toronto manufacturer. The strategists of the Cold War era, preparing to fight vast tank battles on the European plains and not brush-fire guerrilla wars, thought the Caribou too small for their use. Besides, troop carriers were increasingly turboprop, and the Caribou's Pratt & Whitney R-2000s were of Second World War vintage. The aircraft was sent on demonstration tours to the United States, Europe, the Middle East, India, and Australia. Two subsequent tours followed: in 1961 to Latin America and the Caribbean and in 1964 to South East Asia, Africa, and Pakistan. The Indian Air Force did take twenty Caribous to reinforce its Himalayan outposts against Chinese aggression, and as part of a Canadian military aid package, twenty-three were sold on good credit terms to the Kenyans, Ghanians, and Tanzanians. At home the RCAF bought nine Caribous, and also used them on peace-keeping missions; one was destroyed on such a UN mission by a Pakistani F-86 fighter aircraft. The remainder were sold to the Tanzanians.

But the most important customer was the United States military, and fortunately for de Havilland, when the Caribou appeared, the U.S. was about to be involved in Vietnam. The U.S. Army evaluated a number of Caribous in 1960 and placed an order for 22 the following year — eventually buying 164 aircraft. It was first designated as the AC-1, and

DHC Caribou.

when later transferred to the U.S. Air Force, as the C-7. The pilots and infantry loved the aircraft, putting only the Huey helicopter above it. The Caribou's STOL qualities were almost on par with the little Pilatus Porter, yet it could airlift twenty-six troops off a dirt road, mountain strip, or even a paddy field. The only surface they said the Caribou couldn't use was water. Pilots soon learned that it could almost hover in a strong headwind, that it could land in 800 feet or less, and that its landing gear was stressed to a 1,300-feet-per-minute rate of descent touchdown. Its ability to execute steep approaches into jungle clearings meant that it was less exposed to small arms fire.

The Central Intelligence Agency held that the Caribou was custom-made for their clandestine airline Air America. Caribous hauled everything for them: refugees, rice, ammunition, entertainers. Bizarrely, its wide rear doors were used to airdrop so many live animals to displaced indigenous tribes that the Caribou pilots used to say a whole generation of kids grew up thinking pigs fell out of the sky. As they pulled out, the United States military turned over fifty-one Caribous to the Vietnamese National Air Force, and many were captured when Saigon fell in 1975. Increasingly cannibalized, these served in the Communist military until 1998.

From 1964 to 1971, the Royal Australian Air Force also depended heavily on Caribous for its involvement in the Vietnam War, the aircraft being delivered to Saigon directly from the de Havilland plant in Downsview. Three of the RAAF Caribous were destroyed by enemy fire, the remainder surviving to be used in Australian humanitarian and peacekeeping missions in India, Pakistan, Bangladesh, East Timor, and Papua New Guinea, and domestically on flood relief missions in New South Wales in 1990.

When de Havilland ended production in 1973, 307 Caribous had been built. As ubiquitous as the Lockheed Hercules, it had become a staple not only during the Vietnam War but also in humanitarian relief, appearing in the blue and white United Nations livery for decades after. The aircraft even took one final curtain call in 2000 on the television program *Survivor*, when a thirty-six-year-old RAAF Caribou transported the contestants into the outback. Caribous are preserved in museums worldwide; one curiously enough is displayed in Iran as an example of American aggression. In Australia, an air force Caribou was sunk as an artificial reef — something that, when they visualized the uses that their aircraft might be put to, the Downsview engineers certainly didn't think of.

Courtesy of Bombardier

DHC Twin Otter.

DE HAVILLAND CANADA
TWIN OTTER

When Twin Otter pilot John Hemstock flew a BBC television team into the Ethiopian airstrips of Korem and Alamata, he little realized the effect it would have. It was September 1984, and to the starving Ethiopians, the white Mission Aviation Fellowship Twin Otter with its twice-daily flights was their only hope. Later Hemstock recalled, "We were very depressed for the rest of the world did not seem to be aware that the situation existed." Although the aircraft was packed to the ceiling with grain, the crew knew that however many times they landed, they were only deferring death temporarily for the hundreds who waited at the airstrips. But one of the BBC crew was correspondent Michael Buerk, and it was his five-minute news clip that shocked the world out of its complacency, spurring the biggest aid program in history. But until it could arrive, the MAF Twin Otter remained on duty, its three Canadian pilots (Hemstock, Jim McAlpine, and Larry Nicholson) becoming living legends. To the starving Ethiopians, the Twin Otter was the little white bird that brought life. De Havilland Canada could have wished for no better name for its aircraft.

In 1964, de Havilland announced that it was going to build a twin-engine successor to the Otter that would take up to twenty passengers. Pratt & Whitney Canada once more provided the powerplant, the PT-6A-27 turboprop, and on May 20, 1965, Bob Fowler flew the first DHC-6 Twin Otter. Wisely, many of the design features that made the Otter were left intact in the Twin, though a great deal had to be changed. The Otter's double-slotted flap system was kept with the longer wingspan. The wing-struts moved inwards toward the engine nacelles, and the tailwheel was replaced by a steerable nosewheel. It was also the last aircraft that P.C. Garratt, the founder of de Havilland in Canada, would oversee, for he would retire from the company on December 31, 1965, after a career in aviation that had begun in 1915. In 1982, de Havilland brought out two specialized military Twin Otters. The counter-insurgency version had a cabin-mounted machine gun, armour protection, and under-wing hardpoints for rockets or gun pods. The second was the DHC6-300MR maritime version with a chin-mounted radome.

By January 1983 more than eight hundred Twin Otters had been sold, making it the most successful Canadian commercial aircraft export ever — the eight-hundredth was to the Alaska Air

DHC Twin Otter.

National Guard. They were now in service with the military in twelve countries, including Chile, Argentina, and France. Ten Twin Otters supplied to the Peruvian Air Force were float-equipped to be based at Iquitos. In Argentina, the DHC-6 was operated by Lineas Aereas del Estado, a military airline that flew commercially unprofitable routes. Because this included the Argentine base in the Antarctic, they flew in the grey and white Fuerza Aerea Argentina colours with dayglow paint on the noses and tails. In civilian service, Twin Otters have been flown by so many Third World airlines (Royal Nepal, Surinam, Aerotaca, and Haiti Air are typical) that since 1973 they have been hijacked twelve times — including an MAF Twin Otter in Papau New Guinea in August 1999.

In 1970, the Canadian Armed Forces bought eight DHC-6s (designated CC-138s) for transport and SAR duties. They replaced Labrador helicopters at 424 Squadron at CFB Trenton, which provided them for United Nations duty in the Indo-Pakistan war of 1971 and 1972. One of these in United Nations markings was destroyed by the Indian Air Force while on the ground on December 5, 1971. A second Twin Otter with 418 Squadron was lost searching for another SAR aircraft on June 14, 1986. The aircraft hit a mountainside, killing all eight on board. The Air Transport Board ruled that the accident had caused by a freak optical illusion when the colour of the rocks on the mountain, combined with the angle of the sun, made a large ledge impossible to see.

Tuktoyaktuk, Kugluktuk, and Wekweti are about as far away from the villages of Ethiopia as one can get on this planet. But here, too, the residents became familiar with Twin Otters. This is Canada north of 60, an area the size of Quebec, Manitoba, Ontario, and Saskatchewan together. Four of the rugged little transports were later based in Yellowknife, ideal for supporting Rangers and the Canadian Forces Station Alert — the most northern permanent habitation in the world. The men and women of 440 Squadron fly the aircraft year-round, using it on skis, floats, or tundra tires in the summer. Besides National Defence scientists and military personnel to remote areas of the Yukon and Northwest Territories, 440 squadrons crew have also transported governor generals and Prince Charles and his sons. They and their Twin Otters know a Canada that few do.

Courtesy of the DND

DHC Twin Otter.

CF-101 Voodoo.

McDONNELL CF-101 VOODOO

It was meant to be a long-range fighter to escort the B-47 bombers of Strategic Air Command deep into the Soviet Union, but the closest that the McDonnell Voodoo came to Communist airspace was during the Cuban missile crisis and the war in Vietnam. In Canada, it became the mainstay of homeland defence, enabling the country to fulfill its NORAD commitments economically.

In 1947 McDonnell began building the all-weather interceptor Voodoo around two 15,000-pound-thrust Pratt & Whitney J57 turbo jets, and it first flew in 1954. During the Cold War, the Voodoo was used primarily by the U.S. Air Defense Command, configured to intercept Soviet bombers that were expected to come over the polar regions. It was a lucrative aircraft for McDonnell, which churned out 807 Voodoo F-101s in all versions — fighter, interceptor, reconnaissance, and ground support. Able to carry a 3,721-pound nuclear bomb, Voodoos were usually armed with four 20-mm cannon and two AIM-4D Falcons or two AIR-2A nuclear-tipped Genies.

In the mistaken belief that supersonic manned interceptors were obsolete, in 1959, the Diefenbaker government cancelled the Avro Arrow program and replaced it with the surface-to-air missile, the IM-99 Bomarc. When it became clear that a manned interceptor was vital to Canada's air defence, having abandoned its indigenous fighter, Ottawa approached the United States for one. By now, the Voodoos were no longer in their first blush of youth and were about to be relegated to Air National Guard units. Perhaps because the Americans were only too aware of the inadequacies of the Bomarc, a sufficient number was made available to equip the RCAF, thus enabling Canada to meet its requirement to NORAD. Between mid-1961 and 1962, fifty-six F-101Bs were given to Canada, redesignated CF-

101B. The first arrived on November 13, 1961, to 410 Squadron at Uplands, Ottawa. The other RCAF squadrons to operate the Voodoo were 409 at Comox, British Columbia, 414 at North Bay, Ontario, 416 at Bagotville, Quebec, and 425 at Chatham, New Brunswick, with 414 North Bay as the OTU. In 1961, the maintenance contract for the 101s was awarded to Bristol Aerospace, just in time for the first Voodoo crash at Namao in early 1962; Bristol sent its engineers to repair it in the field. On November 29, 1962, the first batch of Voodoos were sent through Bristol for servicing.

Courtesy of the DND

CF-101 Voodoo.

When in 1968, the USAF deactivated seven of its Air Defense squadrons, the surplus market was flooded with Voodoos. As a result, in 1970, in operation Peace Wing, the first CF-101Bs were exchanged for fifty-six of the more modern versions. These had infrared sensors and better fire control systems. In each batch there were ten Voodoos with dual controls, designated CF-101F, to be used as trainers. Now Bristol at Winnipeg became the terminus for newer Voodoos going north from American stocks to RCAF bases and the older Voodoos going south to the Davis Monthan storage base in Arizona. Bristol modified the aircraft, removing certain American components from Northbound 101s and installing them on southbound aircraft from the RCAF. The last northbound Voodoo left Winnipeg on December 20, 1971, the last southbound on January 10,1972, completing operation Peace Wing.

The final Voodoos were exotic creatures. In 1982, 414 Squadron received two Voodoos equipped as electronic countermeasures (ECM) aircraft (CF-101F), one of them the famed Electric Jet loaned from the USAF and the other a trainer. The USAF retired its Voodoos in 1980 and the Canadian Armed Forces in 1985, with the two ECM remaining for another year.

A relic of the Cold War, enough of the Voodoos survive, on pedestals and in museums, to remind us of them.

DASSAULT FALCON

There is a saying in the American aviation industry that goes: "If it's ugly it's British, if it's weird it's French, and if it's ugly and weird, it's Russian." The Anglo-French Concorde apart, if one aircraft proved this wrong, it was the Mystere/Falcon. It came from the nation that invented style. France had already built cars like the Citroen DS and the Facel Vega before Dassault unveiled the cantilever swept, long-wing monoplane at its plant at Merignac. Aviation journalists thought it the best-looking aircraft ever built, and the Falcon could not be mistaken for anything other than French. If was as if Coco Chanel had designed an aircraft.

In 1945, French aircraft manufacturer Marcel Bloch returned to his devastated factory after a year in Buchwald. His crime? Refusing to build aircraft for the Germans. Bloch's brother, Darius-Paul, code-named "char d'assault" (tank), was a hero of the French Resistance. In honour of him, Bloch changed his name to Dassault. At a time when the French Air Force was making do with British-built Vampire jets, Dassault began developing a line of fighter aircraft that vaulted the country into the forefront of aviation. Beginning with the Mystere and then the Mirage, Dassault's fighters equipped air forces from Pakistan to Australia, Israel to Peru, achieving especial prominence in the Six Day War in 1967.

In 1962, Dassault unveiled the Mystere 20, a twin-turbo short- to medium-range transport. What was unusual about it was that the business aircraft used the proven wing of the Mystere, much as the Gates Learjet would later do with a Swiss fighter. Initially powered by Pratt & Whitney JT12A turbojets, General Electric CF700-2D-2 turbo fans were later substituted. The Mystere 20 prototype first flew on May 4, 1963, and when Charles Lindbergh saw it

Falcon.

Courtesy of the DND

Falcon.

perform at the 1963 Paris Airshow, legend has it that he phoned Pan American president Juan Trippe and said, "We've got our plane." The American aviation industry had nothing like the Mystere, and Trippe placed a large order for use in its Business Jets Division as executive transports. Pan American had it configured to seat eight passengers, named it the Fan Jet Falcon (paying Curtiss Wright, the owners of the trademark, to do so), and became its distributor in North America. Besides Pan American, in 1966, Air France and Japan Air Lines also bought Falcons to train their pilots for the new jet airliners. Governments like Spain, France, Indonesia, and Australia ordered the aerodynamically pleasing executive jet as a VIP transport or navigation calibration aircraft. The biggest success came in1976, when the Falcon 20 G, re-engined with Garret ATF3 engines, was chosen by the U.S. Coast Guard as a medium-range surveillance aircraft and designated the HU-25A Guardian. In 1981, the French Navy replaced its Lockheed Neptunes in the Pacific with Guardians.

Canada obtained its Fan Jet Falcons in 1966 as part of a barter deal where the French government bought Canadair CL-2125 water bombers. Beginning with the arrival of the first on May 22, 1967, eight Falcon 20s, designated CC-117, were sent to 412 Squadron to be VIP transports for senior ministers. Expensive to maintain, when the time came to retire them in December 1965, DND hoped that the Department of Transport would take them of its hands. But with its Jetstars and later Bombardier Challengers, DOT had no need of the Falcons, and the last flights of the Canadian Forces transport took place in December 1985, before they were sold.

Three Falcons were also used by 414 (EW) Squadron at CFB North Bay, Ontario, in an electronic warfare role. In March 1987, the last Canadian Forces Falcon was replaced by the Bombardier Challenger. Perhaps in appreciation of the Falcon's pleasing lines, one is on display at the Smithsonian Aerospace Museum, Washington, in Federal Express colours.

Courtesy of Canadair

Canadair Water Bomber.

CANADAIR WATER BOMBER

Mention a SuperScooper and one hopes that it is what owners of large dogs use. But in certain fire-prone parts of the world, a SuperScooper is considered Canada's greatest gift to civilization. To Americans, Croatians, Spaniards, and Venezuelans, the yellow Canadair CL-415T is an angel who dramatically descends from the skies to save their forests, crops, and homes. There are other airtankers and there are other water-scoopers. There are also other search and rescue, fisheries monitoring, and maritime patrol aircraft. But only the CL-415 is multirole, combining all of these duties in one airframe.

Using aircraft for forestry protection isn't new. In 1919, Ellwood Wilson, the chief forester for certain Quebec pulp and paper companies, bought two Curtiss HS-2L flying boats for that purpose and in doing so introduced commercial aviation to Canada. Since then there have been a variety of fire patrol aircraft and waterbombers. In Canada, there are the giant Mars in British Columbia, Cansos in Quebec, and Grumman Avengers in New Brunswick. In the United States, old P-3s and C-130s are rescued from the Davis Monthan scrap yard in Arizona and refurbished as tankers. But what all of these aircraft have in common with the Curtiss HS-2Ls is that they were military surplus, built for another era and use.

In the 1960s, when the Quebec government air service was looking to replace its war surplus Canso water-bombers, Canadair designed a flying boat for that purpose. As flying boats were only being built by the Russians and Japanese, and for naval patrol rather than water bombing, it was an uncertain project. But for a company that had started by manufacturing Cansos in the Second World War, building a flying boat that was also a water bomber presented few difficulties. The aircraft had to have STOL capabilities. It had to be uncomplicated enough to operate under bush conditions, sturdy enough for scooping water in waves up to two metres in height, and capable of carrying up to 1,176 gallons of water or fire retardant to the drop site.

What emerged from Canadair was the CL-215, a twin-engine amphibian powered by Pratt & Whitney R-2800 radial engines. As an amphibian it could operate from lakes, the open sea, and conventional airfields. It would skim a lake's surface and, using two retractable probes below the hull, scoop up water into two tanks in the fuselage and then drop it on the forest fire ... and it would do this with the fastest turnaround time in the business.

Airborne on October 23, 1967, the first CL-215, C-FEU-X, looked as if it had fallen out of a cereal box. With its slab-sided fuselage and cantilever high wing, it was aesthetically unappealing. But the all-metal airframe was designed to be

Canadair Water Bomber.

Courtesy of Bombardier

indestructible, and its high wing would allow it to slow down to drop the water accurately, even in fire-created windstorms. Canadair also developed a lightweight liquid spray system that allowed oil dispersants, pesticides, and chemical foam to be released as well, the aircraft becoming the white knight to environmental disasters. Able to carry up to twenty-six passengers, it was also configured for SAR missions, coastal patrol, and drug interdiction.

The first customer was the Quebec provincial government, which bought fifteen, followed by Manitoba, Ontario, Alberta, the Northwest Territories, and the Yukon. Internationally, France led with fifteen, Greece took twelve, Yugoslavia four, and Italy two. The Spanish government even exchanged Spanish wine for eight CL-215s. Canadair also sent the CL-215 on tours of the Far East, including China; the Thai government bought the flying boat not for firefighting but for coastal patrol.

Bombardier, which owned Canadair, then announced on January 16, 1987, that it would re-engineer the CL-215T with Pratt & Whitney PW123AF turboprop engines. The wings were updated for the more powerful, slimmer engines, and winglets were added for better lateral stability. Power ailerons, pressure refuelling, larger tanks, a more modern cockpit, and a revised electrical system were some of the other improvements. The turboprop waterbomber made its initial flight on June 8, 1989, and hardly had it done so when in October 1991, Canadair launched its CL-415. This aircraft could carry up to 1,350 gallons and had a computer-controlled drop system that allowed the pilot to select the best drop patterns. Customers like Croatia's Ministry of the Interior, which already operated CL-215s, came back for the CL-415 in 1996.

While the usual customers, the provincial governments, the French, Italians, and Greeks bought the 415s, the most publicity came from the leasing of aircraft to the Los Angeles County Fire Department. Here the boxy yellow aircraft became media stars. The firestorms that the county suffered in 1993 were soon an annual event that caused billions of dollars in damage to some of the most expensive real estate in the world. Rather than owning or leasing military surplus tankers, the county negotiated a turnkey lease of two CL-215Ts and crews from the Quebec government. Criticized by the American fire-fighting community, who claimed that county taxpayers were providing jobs for Canadians, the Los Angeles fire chiefs issued a reply. What impressed them, they said, was the Canadair waterbomber's quick response time and ability to rapidly deliver large amounts of fire suppressant repeatedly. This was especially important as with each year the danger of fire in the wildland/urban interface between suburb and scrub increased. With metro areas spilling into the wilderness everywhere, the interface that used to be a safety buffer was disappearing — and only a SuperScooper, with its ability to pick up a full load in twelve seconds and deliver it to the fire quickly, could do interface firefighting.

Bombardier makes faster, more attractive aircraft like the Learjet and Challenger 604, but only its waterbombers have been called the heroes of the skies — and in so many diverse languages. Ellwood Wilson would have been pleased.

Bell CH-135 Huey.

Courtesy of the DND

BELL CH-135 HUEY

The Huey is the Dr. Jekyll and Mr. Hyde of the helicopter world. In some roles it is the warring gunship swooping down on Vietnamese villages (accompanied by Wagnerian music as in the Francis Ford Coppolla epic); in others it is an angel of mercy, an air ambulance omnipresent at earthquake and famine relief sites — the helicopter performs both roles with equal aplomb. Forever associated with the United States' involvement in Vietnam, the Huey became a long-running success story for Bell, operating in sixty-four countries as the backbone of their military, transport, medevac, disaster relief, SAR, and overall utility services.

Until Avco Lycoming developed its T53 turboshaft engine, helicopters were powered by piston engines. Under a joint U.S. Air Force/U.S. Army contract, Lycoming's Dr. Anselm Franz led a team of engineers to develop a light turboshaft with a free power turbine, and this was selected for Bell's new helicopter, the 204, in 1955. This was a simple, basic design with a cabin that allowed for four stretchers. After a comprehensive testing program at the U.S. Army's Fort Rucker, the helicopter, designated the HU-1A Iroquois (H for Helicopter and U for Utility), was put into production and the first was delivered to the U.S. Army in 1959. In 1962 the first HU-1As (called Hueys by the GIs) arrived in Vietnam as helicopter ambulances. Later versions were armed with 7.62-mm machine guns attached to the forward part of the skids and in the rear launching tubes for 2.75-inch rockets. These Hueys were to provide fire cover for the Shawnee transport helicopters, and over the years, with increasingly heavier armament, they developed into the familiar gunships seen in Vietnam. But Bell was already ahead of the Army's need, and by the time the HU-1D flew in 1961 as a troop carrier, it could take twelve troops with its pilot or six stretchers and a medic. Lycoming kept developing more powerful versions of its T55, and Bell adapted the original airframe to cope with it. Bought by nations around the world, the Huey has become associated with Counter Insurgency, its suppressive fire power (sometimes from door-mounted machine guns) feared by guerrillas and drug lords in Sri Lanka, Burma, the Middle East, and Sudan.

On May 1, 1968, the Canadian government approved the development by Bell of the twin-engine UH-1 helicopter with Pratt & Whitney Canada PT6T-2 powerplants. As the federal government had funded much of the research of the

A Special Service Twin Huey offloads its complement of troops. Introduced in 1971, the Twin Huey was primarily used in support of Army field forces. It was flown by the mixed Twin Huey/Kiowa Tactical Helicopter Squadrons of 10 Tactical Air Group (10 TAG).

engine, it was expected that the Canadian Armed Forces would order the Pratt & Whitney-powered UH-1s, and the following year fifty helicopters (designated CUH-1N) were bought, with an option for twenty more. The first of the Twin Hueys were delivered to the armed forces on May 3, 1971, and most Canadians became familiar with them during the FLQ crisis that autumn. The Utility Tactical Transport Helicopter could take eleven fully equipped troops or sling up to 2,900 pounds. Three former fighter squadrons operated the Twin Huey: 422 at CFB Gagetown, 427 at CFB Petawawa, and 430 at CFB Valcartier. When the Chinook and Voyageurs were retired 450 and 424 Squadrons also got Hueys, as did the CFB Goose Bay Rescue Flight in 1993.

The Twin Huey also saw action overseas on United Nations peacekeeping missions in Sinai, Central America, Somalia, and Haiti. Wearing the white UN paint scheme with black nose to reduce glare, the Twin Hueys were equipped with flare dispensing systems to act as decoys for heat-seeking missiles. Replaced by the Bell CH-146 Griffons, the last Huey was retired from 427 Squadron Petawawa on July 2, 1997.

DOUGLAS DC-9

The last aircraft bought for Trans Canada Airlines, the Douglas DC-9 served for more than thirty years with its successor, Air Canada. Like a faithful family retainer, it saw the airline grow up, taking it through several presidents, the freedom of privatization, and changes in logo and livery. When it grew older than most of its crew and passengers, the employees at Air Canada lovingly restored a DC-9 and retired it to the Canada Aviation Museum.

The DC-9 first flew on February 25, 1965, two years before Boeing could get its 737 into the air. Douglas had gotten a jump on its rival but understood that its competition in the short/medium-haul jet category was foreign: the BAC 1-11, the Hawker Siddeley Trident, and the Sud Aviation Caravelle. In the late 1950s, the Europeans were still marginally ahead of the Americans in civil jet aircraft and engines, and for this reason Douglas was the Sud Aviation agent in North America. But when the high operating costs of the Boeing 727 for short routes prompted United Air Lines to buy Caravelles, Douglas was induced to enter the short/medium-haul field itself. It first drew up a smaller version of its DC-8, powered by four Rolls Royce RA50 Avon turbojets. The choice of jet engine worried Douglas — outside of the military, there wasn't much in the United States to choose from. When it came out with the Model 2086, the tried and tested Pratt & Whitney JTF10, designed for military application, was considered. In 1962, Jack McGowan, president of Douglas, and Donald Douglas Jr., son of the founder, travelled to various airlines, including Trans Canada Airlines in Montreal, asking what they could build. The duo was told that from the economic point of view the new Pratt & Whitney JT8D turbofan was more popular. With this, work began on the DC-9 in July 1963, the familiar twin-engine shape emerging in March 1964.

Douglas DC-9.

Trans Canada Airlines, about to change its name to Air Canada, was considering replacements to its Vickers Viscounts and Vanguards. The government-owned airline conducted extensive surveys, studies, and inquiries as to what aircraft it should buy. But being Canada, there were political and cultural implications to any such purchase. Most of the TCA/Air Canada fleet were Vickers, and the Anglophiles among the airline wanted the British dependency to continue. With the airline based in Montreal, French Canadians thought that the Caravelle should at least be considered. When TCA president Gordon McGregor went before a Sessional Parliamentary Committee on December 3, 1963, and spoke his mind about the choice of aircraft he unwittingly made disparaging remarks about the Caravelle. The airline had been struggling with its lack of a bilingual image, and this was all that the French Canadian press and politician Jean Lesage needed to attack it. University of Montreal students rioted on December 14, 1963, shouting "Hang McGregor!" and throwing eggs at the new Place Ville Marie office windows. Never before or since would an aircraft purchase inflame such passions.

In the end the DC-9 won out on practicality, flexibility, and competitiveness. It was the best aircraft for the job — and close to the heart of the federal government. There was to be an industrial spinoff, a joint venture between Douglas and de Havilland to build the wings and part of the airframe at the old Victory Aircraft plant at Malton. Neither Sud Aviation nor British Aircraft Corporation could match that. Air Canada was the first non-American airline to order the DC-9, and between 1963 and February 1973 it would buy a total of fifty in the 14, 15F, 31, and 32 models.

The first DC-9-14 (Fin Number 702), CF-TLC (for Tender Loving Care), landed at Dorval on February 24, 1966, and Air Canada used it on a Montreal-JFK flight on April 6 then put it on the daily Montreal-Winnipeg-Vancouver service on April 24 that year. Not only was it the first aircraft to be delivered in the new Air Canada livery, the DC-9 gave most Canadians their first experience in flying, and jet travel. It truly became the workhorse of the airline. At $4 million

Courtesy of Air Canada

H.W. Seagrim (left), senior vice-president OPS, with gold key presented by Donald Douglas Jr. in Long Beach prior to departure of DC-9 fin 702 CF-TLC for first delivery to Air Canada. January 7, 1966.

each it was a bargain. Self-sufficient on the short routes, it allowed Air Canada to provide jet service to rural communities, and it brought its own air stairs and auxiliary power unit, giving the airline a quick turnaround at busy urban airports. Soon the DC-9-14's seating capacity of 72 proved too small, and on June 12, 1965, 12 DC-9-32s that could seat 115 passengers were ordered. Pilots liked the high-lift wing system of leading edge slats that allowed them to fly slower and used less runway to land. Passengers liked the liftable armrests and that they were fed more quickly: the galleys (the first small airliner to have refrigerators) could

Air Canada Captains Rob Giguere, Hugh Campbell, and Chuck McKinnon deliver the last Air Canada DC-9 to the Canada Aviation Musuem.

accommodate larger meal carts so that the flight attendants didn't have to walk back and forth with their meals. Flight attendants liked the single galley; they could work as a team. And then there was the distinctive DC-9 whine.

As the years went by and more fuel-efficient aircraft appeared, the DC-9's drawbacks became more pronounced: the P & W JT8Ds were noisy — especially for anyone in the windowless back; it was cramped; and the air stairs, control cables, and hydraulic lines were not designed for thirty years of prairie winters — they froze at the worst times.

Finally the day came to take the workhorse out of harness, and in 2001, thirteen were removed, with the remaining five to go by January 2002. Air Canada's last scheduled DC-9 flight (Fin No. 711) was operated by Captain Ken Jones and First Officer Sylvain Boucher on January 18, 2002. The aircraft, older than the crew and most of the passengers, had been in service with Air Canada since 1968 and had logged 81,555 hours, more than any other Air Canada DC-9.

Capt. Hugh Campbell, Senior Director, Flight Operations takes up the story:

In late 2001, Air Canada made a decision to phase out the DC9 fleet. We were also interested in possibly donating aircraft to a Museum and Aviation Colleges. The Canadian Aviation Museum in Rockcliffe was an obvious choice and arrangements were made to have Air Canada donate a DC9 to the Museum. With some research and through experience with the aircraft we determined that a landing in YRO would not be a problem. The result, we had very few current DC9 pilots by September, a few management pilots who had ferried aircraft to the desert during the spring were the most recent, of which I was one. The crew to deliver 711 to YRO was myself as Captain, Captain Rob Giguere (Executive Vice-President Operations and a former experienced DC9 Flight Officer) as in the right seat and Captain Chuck McKinnon (former DC9 Flight Manager) in the jump seat. To prepare for this historic flight we completed a simulator exercise in our DC9 simulator in YYZ. As the aircraft had not flown for almost eight months and the fact that we as a crew now fly B767 or A340 aircraft, we were a little rusty, so we also took 711 up to YMX the day before delivery to YRO for a little practice. As expected our Maintenance group had done an outstanding job on the aircraft and it operated flawlessly with no snags. As for us, the DC9 flies like a dream and after a few circuits, we were all very comfortable with landing in YRO. We practiced a few short field landings and as our takeoff roll was close to 4000 feet, we knew after arrival in YRO there was little chance of getting 711 out of YRO once we landed.

On Saturday, September 21, 2002, we arrived in YUL Operations planning to depart at 09:00. Unfortunately a weak frontal system was just west of YOW and rain was threatening. After consulting with Flight Dispatch, we elected to delay a few hours as a landing on a wet runway at YRO was not acceptable and also after the front went by, we anticipated a bit of a west wind as a bonus to reduce our landing distance. By 11:00, conditions looked good, we filed IFR with plans to cancel and land VFR as we approached YRO. After coordination with Simon at the field and YOW arrival control we were ready to go. In an effort to land at as light a weight as possible we boarded minimum fuel, which included enough to fly from Dorval to Ottawa, conduct a precautionary approach at YRO and if required divert to YOW (total FOB 10,000 lbs). Our take-off weight was 73,000 pounds, giving us an estimated landing weight of 68,000 pounds. Flying time YUL-YRO=45 minutes. As YRO was the shortest runway that any of us could remember an Air Canada DC9 landing on, we planned a precautionary approach to insure we could manoeuvre into a position to touchdown and stop safely. On the first approach I stabilized well out with 50 flap and an approach target speed of 115 knots. As we proceeded in visually, Rob and Chuck monitored altitudes and airspeed so I could visually get the aircraft in position. Our first approach worked out well and after the go around we quickly discussed minor changes for the next approach and final landing of 711. On approach we planned a normal descent to the runway end with my target for touchdown just past the displaced threshold line. (I believe we ended up on the numbers.) As Rob selected reverse thrust, I applied medium braking and we stopped in about 2000 feet, following a smooth touchdown and rollout. Given our practice in YMX the day before, the aircraft performed as expected and our arrival in YRO was no big deal. (Which is exactly how we like to plan all our flights!) We will all miss the DC9, but know that 711 has found a good home!

Sea King.

Courtesy of DND

SIKORSKY SEA KING

Certain aircraft have the misfortune to be politicized at their birth. As with the Avro Arrow and the Anglo-French Concorde, they become symbols of extravagance or parsimony, national pride or disgrace, to feature in a government's electoral platform or as ammunition for parliamentary opposition. In what should have been its Golden Years, the Sea King helicopter suffered such humiliation. It became the septuagenarian that wasn't allowed to retire.

In May 1964, when the Royal Canadian Navy bought Sikorsky Sea Kings to replace its HO4S-3s, no one thought that four decades later they would still be flying. At that time, Canada pioneered the use of helicopters on warships, effectively doubling their anti-submarine and surveillance effectiveness. The Sea King had a single step boat hull incorporated into its all-metal, semi-monocoque fuselage. Two stabilizing sponsons or floats mounted on each side were also stowage for the twin wheel main landing gear. The tail and rotor blades could be folded to facilitate storage in specially constructed shipboard hangars, and the Canadian-invented Beartrap allowed it to land on ships rolling in the high seas. Unlike its predecessor, the Sea King was amphibious, and had an all-weather, day-night capability. Its active dipping sonar, installed later, was optimized for the localization of diesel/electric submarines in shallow waters, which provided ideal havens for submarines to lurk undetected. It was really the latest in helicopter technology.

This Sikorsky model was also bought by the British, Australian, Peruvian, German, Indian, Pakistani, and Italian navies and the Egyptian and Royal Norwegian air forces. It was also built by Westland in the United Kingdom and Augusta in Italy. The airframe was a good weapons platform — even Exocet missiles were carried by some operators — and that it could be quickly modified was evident during the Falklands War, when Royal Navy Sea Kings were pressed into service as airborne early warning systems, distinguishable by the radar kettle drums mounted on their starboard side. Actually, the best advertisement for the Sea King model occurred in 1961, when it was chosen by the U.S. Marines for the Presidential Flight — a role it continues in today.

Sea King.

The Royal Canadian Navy (RCN) initially operated forty-one Sea Kings, the first arriving at RCNAS Shearwater, Nova Scotia, on May 24, 1963. United Aircraft Ltd. in Montreal assembled the components of all except the first four, which came intact from Sikorsky, and all were assigned to operate from the aircraft carrier HMCS *Bonaventure* and from the helicopter destroyer escorts (DDHs). With its Trackers and a HO4S plane guard, the *Bonaventure* normally embarked four to six Sea Kings. The smaller class DDHs carried one Sea King, while the larger Tribal Class DDHs had two. When HMCS *Bonaventure* was paid off in 1969, the helicopter anti-submarine squadron was divided into two operational squadrons, and the former RCAF squadrons HS 423 and HS 443 were reactivated on September 3, 1974. The *Bonaventure* Sea Kings were then based at Shearwater, Nova Scotia, and painted in the standard Canadian Forces overall gray-green paint scheme. Between July 1987 and May 1989, 443 HS was moved to the West Coast at Patricia Bay.

On both coasts, the Sea Kings (now designated CH-124A) laboured through the decades, seeing unification of the forces and enduring countless refits and changes in their engines and electronics. For although the oldest aircraft in the Canadian military, the Sea King was also the busiest, participating in international and domestic operations that included Op Friction in the Persian Gulf, Op Deliverance in Somalia, Op Sharp Guard in the former Yugoslavia, Op Forward Action in Haiti, and Op Determination in the Arabian Sea. Closer to home, Sea Kings took part in Op Assistance in the Winnipeg floods, and Op Persistence in support of the Swissair 111 crash off Nova Scotia. In August 1990, when Iraq invaded Kuwait, three naval ships were sent to the Persian Gulf as part of Canada's commitment to a United Nations multinational force to restore peace. The DDH, HMCS *Athabaskan,* embarked a two–Sea King detachment, and HMCS *Protecteur* embarked a three–Sea King detachment, both from Shearwater's

HS 423 Squadron. The five Sea Kings were the first Canadian combat aircraft to participate in the Persian Gulf action, dubbed Operation Friction. Operationally, they were tasked to interdict unauthorized merchant vessels in the Gulf, protect the vital logistic sea lines, and search for mines.

The Gulf War showed up the Sea Kings' age, particularly when it came to self-defence. For this campaign eight were equipped with a temporary anti-missile defensive system, but it was removed when they returned home. The federal government was reluctant to spend the $2.8 million needed for what was basic NATO helicopter defence, reasoning that the whole fleet of Sea King survivors was long overdue for replacement.

Throughout the 1990s, despite embarrassing incidents, federal ministers kept insisting that the Sea Kings were safe, pouring $80 million to upgrade the aircraft — and senior military officers joined in the chorus. Armed with such ministerial logic, the Sea Kings were sent unprotected into the deadly guerrilla warfare of Somalia in 1993 (even as two U.S. Blackhawk helicopters were shot down) and into Haiti in 1994. Pleas to Prime Minister Jean Chretien's government to have an anti-missile system installed, especially after it cancelled the EH-101 Sea King replacement contract in 1993, were disregarded. The opposition parties and media accused the Prime Minister of personally thwarting the replacement program. In January 2001, DND warned about the difficulty of continuing upgrades: the challenge of maintaining a 1950s design not only resides with the assurance of structural integrity but also suggests a different approach when new technologies are harnessed to support older designs.

Sea King delivering supplies to Canadian troops in Somalia.

It took the American campaign against terrorism after September 11, 2001, and the prospect of the Sea Kings going once more to the Middle East for action to effect change. The aged helicopters got their missile defence when the Helicopter Operational Test and Evaluation Facility (HOTEF) and technicians and armourers from 12 Wing Shearwater deployed to Valcartier, Quebec, for trials on the Sea King self-defence suite (SDS), officially dubbed XENA 2001. Composed of a missile-approach warning system and a system for automatically launching chaff, or flares, to distract or confuse incoming threats, it also includes an infrared jammer to widen the spectrum of protection afforded to aircrew — just in time for Operation Apollo.

Patching in hi-tech equipment wasn't the solution. It didn't take the Canadian media long to find out that in this Persian Gulf deployment a Sea King on HMCS *Charlottetown* was taking up valuable hangar space because it was usually out of commission and a third of its missions were cancelled. Closer to home was the embarrassing cross-Canada journey of a Sea King that left Esquimalt, British Columbia, on June 21, 2002, to arrive at Halifax, Nova Scotia, fourteen days later after forced landings in Calgary, Alberta; Duluth, Minnesota; and Ewen, Michigan. By February 26, 2002, when the Sea King on the Mid-East-bound HMCS *Iroquois* lost power during takeoff and smashed onto the deck, the whole country knew the helicopter's miserable history by heart. Of the forty-one acquired in 1963, twelve had crashed, twenty-three had been damaged but repaired, and eight had been severely damaged but also repaired. Ten deaths had resulted from Sea King accidents, 6 people had suffered major injuries in accidents, and 105 had suffered minor injuries. That the men and women of the armed had been able to keep the twenty-nine remaining flying for forty years was a testament to their ingenuity — and courage.

The farflung military operations that the Sea King has been thrust into didn't mean that its SAR duties had been neglected. The latest of the many rescues that the helicopter featured in took place on March 23, 2000. That night, three Sea King Airborne Electronic Sensor Operators, Sergeant Philip Trevor, Sergeant Fred Williams, and Sergeant Tony Thomas of 12 Wing Shearwater volunteered to rescue twelve surviving sailors from a bulk freighter that had sunk off the coast of Nova Scotia, forcing its crew into the stormy seas. Although not trained in search and rescue operations, the men volunteered to be lowered from two Sea King helicopters hovering in the dark above the six-metre swells. Sergeant Thomas was lowered into the debris and oil-covered ocean and tried to reach one victim's side until he suffered a serious back injury when hit by a large wave. When he was hoisted back up, Sergeant Williams took his place and continued the search for survivors. For more than an hour, he was tossed around by high winds while moving into position to recover victims. As he reached a group of three clinging to one another, the panicked men forced him underwater several times before he was able to isolate and airlift each of them to safety. He then was lowered into the sea once again to perform one last and difficult rescue of another sailor. Meanwhile, from the other helicopter, Sergeant Trevor, who knows how to swim a bit, was lowered eight times into the raging seas. Although he was repeatedly pushed underwater and dragged throughout the ordeal and nearly drowned from salt water and oil ingestion, he persevered despite weakening strength until he had recovered six men from two life-rafts and the last two survivors still in the freezing water. Sergeant Trevor remained modest about the events of that night, saying, "All in all it was a pretty busy night with some gnarly surf. I'm just glad that we were able to make a difference. I didn't really have time to think about what we were doing. We just did it." Sergeant Trevor paid tribute to all the crew members involved in the rescue, saying that the two Sea King pilots, Captains Bill Fielding and Perry Comeau, "did some pretty amazing stick-handling to keep the Sea Kings steady and Captain Dave Beerman, the tactical navigator, did some fantastic conning get the aircraft into position over the survivors." The technicians on the navy ship nearby outfitted the second Sea King in record time to remove some equipment in order to lighten the load in preparation for airlifting the fourteen survivors. For acts of conspicuous courage in circumstances of great peril, the three sergeants were awarded the Star of Courage.

For the forty-year-old Sea King, it was just another day at the office.

DHC Dash 8.

Courtesy of Bombardier

DE HAVILLAND CANADA DASH 8

When the Dash 8-400 made its first flight on January 30, 1998, the test pilot commented, "It was flawless, I can't say that enough times. This is a delightful airplane to fly." The relief that de Havilland and its 5,500 employees felt reverberated as far as Montreal (the headquarters of its parent, Bombardier), Queens Park, and Ottawa. Since1980, when the original Dash 8 design was made public, the aircraft had endured a sad childhood and been passed among several foster homes before it could bloom.

In 1978, de Havilland Canada conducted an extensive survey to figure out what the customer wanted. They discovered that regional and commuter carriers asked for a short-haul, short takeoff and landing, fuel-efficient turbo-prop airliner that would be allowed to use inner-city airports because it was quiet. In August 1980, with these surveys in hand, de Havilland announced that it was building the Dash 8, a 36-passenger turboprop with a range of 1,100 miles. Four prototypes were built; the first, C-GNDK, made its initial flight on June 20, 1983, with the others following in November 1983 and January 1984. Two Dash 8-100s (designated CC-142) were bought by the Canadian Forces to replace the two Dash 7s in service with 412 Squadron in Lahr, Germany. Later, DND bought four Dash 8s (CT-142) to be used as navigation trainers for 402 Squadron, CFB Winnipeg, the first arriving in 1989 and the last in 1991 — all highly visible with their protruding Gonzo noses. Two CC-142 transports were added later.

Initial sales were made to NorOntair, the Ontario government's northern air company, and de Havilland looked to sell Dash 8s to second-and third-tier airlines. The hub-and-spoke era in air travel had just begun, and with its STOL features, the aircraft was perfect to feed rural passengers into the large airports. Noting this, the aeronautical giant Boeing thought it a good fit for its aircraft family and in 1986 bought de Havilland from Ottawa. But while de Havilland had some brilliant designs in the Twin Otter and Dash 7, its old-fashioned Downsview plant and militant work force left something to be desired. Boeing expected to pour some money into modernizing the company, but after $400 million (and one bitter ten-week strike), the Seattle company was ready to unload it.

By now the Dash 8-300 was flying, seating between fifty and fifty-six passengers and offering possibilities to be stretched further. There was a market for the Dash series — espe-

DHC Dash 8.

cially in crowded Europe where air taxi and commuter airlines like Tyrolean, Wideroe, Trans Travel, Augsburg Airways, Cirrus Airlines, and Rheintalflug snapped up the Dash 8s of the 100, 200, and 300 series. Larger airlines like U.S. Airways and British Airways (with its subsidiary BA CitiExpress) also added Dash 8s to bring passengers into their hub airports. The U.S. military and Australian Customs did so as well, but more unusual was the purchase of a Dash 8-200 by Laser Airborne Depth Sounder of Adelaide, Australia, to be an airborne platform for a hydrographic surveying of shallow waters.

But none of these sales were enough to help de Havilland's bottom line, and in 1990, when the European consortium Avions de Transport Regional, themselves successful feeder aircraft builders, announced that they were interested in taking the company off Boeing's hands, it looked as if the Dash 8 was going to yet another foster home. Rather than allow this to happen, Bob Rae, the New Democrat Premier of Ontario, and Michael Wilson, the Federal Minister of Industry, offered de Havilland (with enough financial enticements) to Bombardier. By now the company had shrunk to 2,740 personnel and was struggling to put 32 aircraft out annually. In March 1992, Bombardier organized it so that so that three years later at the 1995 Paris airshow, it could announce the Dash 8-400 program. To prime the pump for the new aircraft, the federal government's industry minister, John Manley, announced on December 17, 1996, that Technology Partnerships Canada would loan de Havilland $57 million in the form of a royalty payment on the Dash 8-400.

DHC Dash 8.

The new Series 400 emerged as a faster, seventy-passenger version of the original Dash 8 and the ultimate stretch of the basic design. By the time the first flight of the 400 had taken place de Havilland had sold 63 — one third of the number it needed to break even. When the historic first flight took place on Saturday morning, January 30, 1998, at 11:00 A.M., it was the culmination of almost two decades of schemes, loans, and hope for the aircraft. The test pilot flew a few cautious laps over Downsview before heading in a slow, level flight for Trenton. Soon it was doing steep turns, sharp climbs, and even sideslips in the air — actions unusual for a first flight.

"The pilot must be really confident," one of the test engineers said. "They rarely do that on the first flight."

Fokker F28.

Courtesy of Canadian Airlines

FOKKER F28

The Dutch aircraft manufacturer, the Canadian airline, and the bird logo had come together before — more than half a century earlier. James Richardson would have approved of the restoration of the Canada goose logo by Canadian Airlines in 1999, as he would have been familiar with the airline's economic problems. He would also have known the Flying Dutchman, Anthony Fokker, the original manufacturer of the aircraft that Canadian Regional Airlines and InterCanadian were using. After all, in 1927 he had introduced Fokker aircraft to Canada to begin Western Canada Airways with, even planning to build them in Winnipeg.

The Fokker F28 1000 Fellowship that Canadian Airlines flew was a link to the successful F27 Friendship, the high-wing, turboprop aircraft that had single-handedly revived Fokker's balance sheet after the Second World War. In 1965, with a loan from the Dutch government and collaboration with Short Brothers in the U.K. and MBB in Germany, Fokker embarked on building a short/medium jet transport. The twin jet made its first flight from the Schipol factory on May 9, 1967, and the first production aircraft was delivered to the German charter airline LTU on February 24, 1969. Fokker ran an employee competition to name the jet, and Fellowship was the winning entry. Several countries bought the F28 for official use, and it became the preferred presidential and VIP transport for Malaysia, Philippines, Nigeria, the Ivory Coast, Gabon, Ghana, and Colombia.

While the early model F28 was not as great a success as Fokker would have liked, it served as a basic design from which the Dutch would build for themselves a niche as a supplier to regional airlines. In 1980, when the company saw that domestic airlines wanted a wide-body, short-haul airliner with six abreast seating, it stretched the F28, added a wing that could give it a cruise speed of Mach 0.75, and then searched for new engines for it. McDonnell Douglas was also looking for a 150-seat, short-haul transport, and until 1982, Fokker collaborated with the American company. When this did not work out, Fokker, once more independent, revived the F28, stretched it, and discovered the Tay 620-15 turbofan engine, a variant of the model that Rolls Royce was building for Gulfstream. One of the best aero engines ever built, the Tay (now called the Rolls Royce RB 183-2, Mk555-15P) gave the F28 an increased

Fokker F28.

maximum coefficient that reduced stall speed, improving takeoff performance. The design had a new cabin layout that provided for 107 seats, 5 abreast (aisle off-set to port), with a range that varied between 2,483 and 2,965 kilometres.

One of the first airlines to buy the new F28-1000 was Wardair Canada, which ordered twenty-four but went out of business before any were delivered. In 1973, Transair became the first Canadian airline to use F28s, operating two of them, C-FTAV Fort Resolution and C-FTAY Fort Prince of Wales. The airline made history when Rosella Bjornson was hired as First Officer on the F28, becoming not only the first female airline pilot in Canada but also the first female in North America to fly a commercial jet. When Transair was bought by Pacific Western Airlines in 1977, its aircraft were gradually sold off, the F28s going to Air Nuigini.

With more orders and financing from the Dutch government, Fokker launched the F28 Mark 1000 short fuselage version, and several feeder airlines bought these sprightly little airliners that seated 55, 60, or 65 passengers, notably Canadian Regional and Inter-Canadian. In January 1989, the latter had leased seven F28-1000s through the International Lease and Finance Corp, and Canadian Regional purchased more F28s in1999, many of which would still be in the fleet when the airline redesigned its logo.

Fokker F28s were also used by other small operators in Canada: Norcanair, Air Ontario, Air Niagara, Atlantic Island Airways, and Canadian North. The only F28 crash in Canada took place on October 25, 1988, when Air Ontario F28-1000 C-FONF took off from Dryden, Ontario, and crashed into the trees, killing twenty-one passengers and three crew. By August 2000, Canadian Regional Airlines operated twenty-eight F28 Mark 1000s. Passengers found them cramped, and when it merged with Canadian, Air Canada considered the aircraft uncompetitive and disposed of them as soon as possible. By then, the Fokker company had gone out of business, the F28 becoming the last of the Flying Dutchmen.

DOUGLAS DC-10

"This aircraft has been a workhorse for CDN through the years," said the airline's fleet manager, Bob Comack. "It has reflected a global presence for CDN on every continent except Africa. It may be an aging aircraft type, but it has and still does serve us well." It was February 2000, and the McDonnell-Douglas DC-10 workhorse was about to head for the pasture.

In 1968, with Boeing poised to capture the jumbo jet market, both McDonnell Douglas and Lockheed sought to design their own such jets. As was the style of the day, both chose to put an engine in the tail — literally. As Boeing did with its 707, to supplement research and development costs, in 1977 Douglas sold the DC-10 concept to the United States Air Force as a refuelling tanker, the KC-10 Extender. The first Extender flew on April 16, 1980, and subsequent orders to a total of thirty-six KC-10s kept the Douglas line at Long Beach open. Cracking the commercial markets proved harder — Lockheeds L-1011 Tristar was larger, cheaper, and more popular, especially with regard to the third tail mounted engine. Both United Airlines and American began flying the DC-10 in August 1971, but several other airlines balked. For one thing, airline maintenance preferred working on the Tristar's S-shaped inlet to the banjo-shaped DC-10. Eastern Airlines, British Airways, Delta, Nippon, Cathay Pacific, and Air Canada snapped the Tristar up, but it was a decision that they may have regretted later when Rolls Royce, the manufacturer of the Tristar's RB 211 engine, went through a financial crisis. For unlike the DC-10 tail, which could take various engines, only the RB-211 could be installed in the Tristar.

James McDonnell, the new owner of Douglas, chose compete with Boeing and its 747s and enter the long-range market with the DC-10-30s and 40s. With their General Electric CF6-50 engines, extended wingspan, and all-up

Douglas DC-10.

weight of 572,000 pounds, the 30 and 40 models were fuel-efficient over long runs. The DC-10-30 appealed to airlines with a global reach, like Singapore Airlines, KLM, United Air Lines, Northwest, and CP Air — and McDonnell made the DC-10-30 ERs (for Extended Range) available before the long-range Lockheed L-1011-8.4. The ERs, with their heavier gross takeoff weight and additional fuel capacity, were distinguished by the extra landing gear in the belly.

The DC-10 earned its poor reputation as the Douglas Death Cruiser or Death Star through three accidents early in its career: the cargo door blowing open on a United Air Lines DC-10 in 1971; the decompression of the Turkish Airlines DC-10 in 1974 that killed all on board; and the American Airlines DC-10 that lost an engine on takeoff from Chicago in 1979. These were the DC-10-10s and not the DC-10-30s that Canadian Airlines used, and staff reminded passengers about this difference. But to the public, all 10s were tarred with the same brush.

CP Air exchanged four of their 747s with Pakistan International Airlines for four of their DC-10-30s. It then leased the first two to Varig, and it was the third DC-10-30 that entered service in November 1979 and was the first in Canadian Airlines colours. At one time Canadian would have thirteen DC-10s in its fleet, cross leasing with United Airlines for three 10-10s but keeping the U.S. registrations. They would be used on the Toronto-Tokyo, Vancouver-Auckland, Santiago-Shanghai, or the high-density Vancouver-Calgary-Toronto runs. The 10s became Canadian Airlines' own aircraft — for apart from the three aircraft owned by Wardair, which it absorbed, only Canadian, as CP Air and Canadian Pacific, flew the 10 in Canada.

Canadian Airlines also appreciated them for their very strong airframes. On October 19, 1995, a Canadian Airlines DC-10-30ER was taking off from Vancouver bound for Taipei when it experienced engine failure at the critical V1 speed. The aircraft departed the runway at forty knots, and although the nosewheel collapsed and there was substantial airframe damage, there were no injuries, and within a few months it was back in service.

To the young pilots with their glass cockpits, the DC-10 was the Douglas Diesel. But the older ones appreciated the roomy cockpit of what became a three-pilot aircraft, calling it the Douglas Diner because the second officers seat had its own table, allowing for more elegant dining than balancing a meal tray on one's lap.

Through the years and the good and bad times, Canadian Airlines' superior maintenance kept their DC-10 fleet in perfect condition. Through lease-back deals, none of the DC-10s were actually owned by the airline — four, for example, were leased from Frankfurt and Nagoya banks — but as the only Canadian operator of the aircraft, the airline became synonymous with them. In 1995, after its employees had bought a 25 per cent equity stake in the airline and their labour unions agreed to reduced salary demands, Canadian Airlines ran an ad campaign that made this point. Television ads began with, "If your good name were riding on something ..." followed by a DC-10 festooned with signatures of the Canadian employees who presumably had worked on it. Never mind that the signatures were actually decals (the employees were dissuaded from using pens to sign their names on the aircraft) and that the Calgary head office wanted a more impressive 747-400 for the ad. The choice was an appropriate recognition for both the hardworking employees and the DC-10s of the airline.

The two DC-10s used, 911 and 912, were the signature aircraft, the former known as The Spirit of Canada and the latter as The Pride of Canada, the names chosen by the employees. "The idea," said a longtime employee, "was to recognize that we were an employee-driven airline." Each aircraft held six hundred signatures in larger-than-life size, and it was a source of pride for a Canadian Airlines employee battered after years of salary reductions to point to his or her name in front of the third engine. As a fortunate staff member said, "It was an

Top and left: Douglas DC-10.

honour to have my signature on the plane, but there were over thirteen thousand other employees who were also worthy of having their names on them as well." Both DC-10s became the media stars of the airline, and there were passengers who asked if they could be booked to fly on them.

Struggling for profitability, in 1999 the airline down-sized capacity on the London, Honolulu, and Japan routes, replacing the DC-10s with Boeing 767s. The new fleet renewal plan had made them too costly to maintain. On November 1, four of the aircraft, 911, 912, 903, and 904, were returned to their lessors, the Wells Fargo Bank in Salt Lake City and Finova Capital Corp in Scottsdale, Arizona, to raise financing to buy two brand new 767-300ERs and one used leased A320. The last DC-10s were phased out of Canadian Airlines by March 2000, just as the merger with Air Canada unfolded. It was as if both events coincided to mark the end of an era in Canadian history.

Griffon.

Courtesy of the DND

BELL CH-146 GRIFFON

The gryphon (also written as griffon or griffin) is a mythical creature that has the head and wings of an eagle and the body of a lion. Usually depicted as heroic symbols, gryphons were well known for their speed, ability to fly, and having the sight of an eagle and the strength and courage of a lion. In Roman art, gryphons pulled the chariot of Nemesis, the goddess of justice and revenge. The Canadian government hoped that its helicopter so named would be all that — and more.

In civilian life a Bell 412-horsepower, the CH-146 Griffon was selected by the Canadian Forces to replace three of its more dated helicopters, the Twin Huey, Kiowa, and the Iroquois, thus reducing training, support, and maintenance but assuming all of their roles. Often mistaken for the Twin Huey, except for its four rotor blades against the Huey's two, the Griffon was built at Bell Helicopter Textron at Mirabel, north of Montreal. The procurement of one hundred airframes began in 1995 and ended in 1998.

Its primary mission was armed support — it had C9 (7.62) machine guns mounted in the open doorway and could carry up to a dozen lightly armed troops or eight fully armed. The crew of two pilots and one flight engineer had also to be skilled in medevac and SAR. The best thing about the Griffon, said Minister of Defence Art Eggleton on Jnauary 30, 1998, when his department took delivery of the last of the helicopters, was that it was a commercial off-the-shelf product that had been customized to meet the needs of the Canadian Forces. Not only did this cut development costs and reduce the project's duration, but the Griffon could be fitted with any one of thirty mission kits. FLIR (Forward Looking Infra Red), a SAR winch, and a powerful searchlight for night visibility could be easily fitted to its airframe. They were based in Goose Bay, Newfoundland; Gagetown, New Brunswick; Bagotville, St-Hubert, and Valcartier, Quebec; Petawawa and Borden, Ontario; and Cold Lake and Edmonton, Alberta. By far the largest distributions of Griffons were to 408 (Edmonton), 427 (Petawawa), 430 (Valcartier), and 400 (Borden) tactical helicopter squadrons; the combat support squadrons at Cold Lake, Goose Bay, and Bagotville also got three or four each.

Griffon.

As Canada's workhorse helicopter, the Griffon was soon involved in humanitarian relief efforts at home — in 1996 the Saguenay floods, in 1997 the Manitoba floods and the Ice Storm. Like the Hueys before them, they also earned their peacekeeping stripes in United Nations operations, becoming prominent in 1997 as part of the UN Support Mission in Haiti, in Kosovo with NATO and Bosnia in the stabilization force, and in 1998 in the Honduras. The Griffon got worldwide media coverage when it starred at the Kananaskis G8 Summit in the 2002, providing security and transport for heads of state.

Given the dangerous operations, both natural and manmade, that the CH-146 has been involved in, it is a wonder that to date so few have crashed. In November 1996, a Griffon from 444 Squadron, Goose Bay, crashed into the icy waters of Northern Labrador where it remains today. On July 18, 2002, another Goose Bay Griffon was returning from an aborted SAR mission. The crew, Captain Colin Sonoski, Captain Juli-Ann Mackenzie, and Flight Engineer Mario Michaud, were looking for an overdue boat near Hopedale. It suddenly lost contact with its base, and the downed helicopter was later found forty-five kilometres northwest of Goose Bay by another Griffon crew. Both captains were killed and the flight engineer was hospitalized.

Beginning on July 26, 2002, the tail rotors on the Canadian Air Force's fleet of CH-146 Griffon helicopters were subjected to a detailed visual inspection before their next flight as a precautionary measure in light of recent findings by the accident investigation team. The inspection was after technical and flight safety experts assessed that the July 18 crash was probably due to a tail rotor failure. Technicians paid particular attention to a specific section near the tip of each tail rotor to look for nicks, scratches, or dents that were larger than previously established engineering parameters. The inspection was similar to work normally conducted every twenty-five flying hours. In November 2002, all Griffons were grounded when cracks were discovered in the root of the main rotor blades of several. Maintenance technicians applied epoxy to the cracks and allowed twenty-four hours for drying.

By mid 2004, the Canadian Forces is scheduled to reduce its number of Griffons from ninety-eight to eighty-five — the thirteen aircraft to come from the tactical and reserve squadrons rather than the combat support. The Royal New Zealand Air Force was said to have expressed an interest in them. In legend, gryphons were to known find gold and lure men to it, devouring them as they came. One can hope that the future of the helicopters so named will not be as grievous.

Boeing 767.

Courtesy of Canadian Airlines.

BOEING 767

An aircraft is a vehicle capable of supporting itself aerodynamically and economically at the same time. So goes the mantra of every designer since Bill Stout launched the Ford Trimotor. Achieving the right balance between economics and aerodynamics was what made a best-selling aircraft. With its 767, Boeing seems to have accomplished it.

When on September 26, 1981, the Boeing Model 767-200 first took to the air at Paine Field, Everett, the company was already locked into a race with Aerospatiale. That day, through no coincidence, the first A300 with forward facing crew cockpit avionics, which made a two-man crew possible, also flew. Boeing had waited for six years after the initial flight of the first Airbus to propose a development of its 707 family — and by then it had lost a full 30 percent of the commercial market. Through its next generation of aircraft, the 757, 767, and 777, it intended to win its lead back.

The 767 first entered commercial service with United Airlines on September 8, 1982, the launch airline for both the 767 and 777, and within a decade became the most widely used airliner across the Atlantic. By April 30, 1999, a total of 746 Boeing 767s had been delivered, the last to the Russian airline Aeroflot. More than two-thirds of the number were 767-300 ER (extended range) versions. In February 1982, Boeing announced that it was going to build an extended version of the 200 series. By inserting two fuselage plugs, one forward and one behind the wing section, the basic model had been stretched 6.42 metres. The first 300 flew on January 30, 1986, and was awarded its certification that September. The twin jet, two-aisle plane with seating capacity of 218 passengers in three classes, or 269 in two classes, was capable of flying up to 7,080 miles in one stretch, could reach a maximum altitude of 35,200 feet (10,725 metres), and had a cruising speed of Mach 0.80. This made possible such nonstop flights as New York to Beirut, London to Bombay, and Tokyo to Sydney. The planes engines could be Pratt and Whitney PW4000s, Rolls Royce RB211-524s, or General Electric CF6-80C2Bs — all designed to be able to operate individually, if one sustains damage. Boeing quoted a price of US $97.5 million per jet, and by far the biggest users were Delta and American Airlines. One of the Boeings

Boeing 767.

Courtesy of Canadian Airlines.

slated for Delta Airlines figured in the world of espionage when it was revealed that the 767-300ER had been bought by the Chinese government as the personal aircraft of President Jiang Zemin. It arrived in Beijing equipped with presidential beds, plush toilets — and bugs. So many listening devices were discovered hidden in the 767's cabin that the Chinese general responsible for presidential security was fired.

Like the 707 before it, the Boeing 767 has been adapted to military uses. It is the platform for the 767 AWACS Airborne Warning and Control System in service with the Japanese Defense Agency, and the military Tanker Transport aircraft for the Italian Air Force. Boeing's biggest coup will be if it is bought as a replacement for the KC-135 Stratotanker by the U.S. Air Force.

As with most new aircraft, the 767 had its share of ill fortune. Besides arsonists, two highjackings, flight control problems, and one loss of simultaneous power in both engines, 767s have suffered ingestion of birds at Logan Airport, Boston, and Mombasa, where flamingos disabled a Kenya Airways 767 on takeoff. An American Airlines 767 waiting for maintenance rolled through the Los Angeles Airport fence and came to a stop on a side street. In March 2000, Boeing notified all its 767-300 customers that parts of the wing structure near the central fuel tank were susceptible to cracking. A year later, Ansett Australia grounded its fleet of ten 767s, when cracks were found in the engine pylon mountings. The causes of the two most tragic crashes are still controversial: in May 1991, the Lauda Air 767-300ER crashed in Thailand because one of its engine thrust reversers accidentally deployed during a climb. On October 31, 1999, an Egypt Air 767-300ER off the east coast of United States plunged out of control, killing all

Courtesy of Canadian Airlines.

on board. Although Egyptair asked that the pitch control on the 767 be investigated, blaming it for the loss, both Boeing and the FAA insisted that subsequent findings and new checks showed that this was not the case. The National Transportation Safety Board determined on March 13, 2002, that the probable cause of the EgyptAir flight 990 accident was the airplane's departure from normal cruise flight as a result of the relief first officer's flight control inputs. The reason for the relief first officer's actions was not determined.

Air Canada initially operated twenty-one 767-200s, one of which, AC FIN 604 C- GAUN, would go down in history as the Gimli Glider. By 2002, these had been augmented by thirty-five 767-300-ERs, and one of them, AC FIN 655 C-GHML, was involved in the aircraft's only mishap to date. On May 13, 2002,on final approach into Toronto's Pearson International Airport, the flight crew received an aft cargo bay fire warning. They activated the cargo bay fire extinguishers and declared an emergency, landing safely. Investigators later confirmed that the fire was a direct result of an electrical failure of a heater tape used to prevent water lines from freezing.

Into the sunset: the Canadian Airlines Boeing 767 in Multimark livery.

Canadian Airlines inherited its first two 767-200s from Pacific Western Airlines. C-GPWA and C-GPWB had been acquired in March 1983, directly from Boeing. Canadian would purchase its own 767-300s in March 1989, disposing them in December 1997, when they would fly for Varig. The airline would eventually use eleven 767-300ERs, putting them on the ten-hour Toronto–Sao Paulo route, as well as to Honolulu, Tokyo, and Beijing. Coping with fiscal uncertainty after 1995, Canadian put off new aircraft leasing, and it would not be until September 17, 1998, that it was able to take possession of the first of two Boeing 767-300ERs from Asiana Airlines. To replace its DC10 fleet, two brand new B767-300ERs were also being leased from GE Capital Aviation Services.

The 767s featured extensively in Canadians campaign to go after the business executives and to lure the high paying road warriors. It introduced initiatives like new Millennium J seats, and power to the seats on the 767s and the

Chiefs Conclave. But the most potent and historic of all was what occurred on January 13, 1999, with the launch of Proud Wings. Although a Boeing 747-400 was used for the launch, as none of the 767s could be taken out of service, it was never only about painting a stylized goose logo onto a plane. Canadian Airlines was then in desperate straits and it made sense to put on a brave face and troll for potential investors. Proud Wings gave both Canadian's embattled employees and its loyal customers a much-needed boost. The artwork had to be translated into paint masks to fit the curves of the fuselage, not only of the 747 but of the DC10-30, 767, 737, and Fokker 28 as well. There were initial fears that because the goose did not have an eye it would be perceived by Asian customers as flying blind, and Canadian went to some pains to point out that the bird insignias used by Japan Air Lines and Singapore Airlines were eyeless as well.

If the federal government's Team Canada flights always used Canadian's Boeing 747-400s, the 767 would have its day in international diplomacy as well. In May 1999, the Proud Wings 767 would carry a high profile Quebec government trade delegation to Mexico City. When Canadian merged with its rival Air Canada later that year, the 767s would be reborn once more, this time in red and white maple leaf livery, which while functional could never equal the beauty of Proud Wings.

CH-149 CORMORANT

When the first two Cormorants arrived at CFB Comox on October 11, 2001, they ended not only a 6,500-nautical-mile, two-week voyage from Verigate, Italy, but also a long wait. The aircraft had flown from Verigate, stopping for fuel in Britain, Iceland, and Greenland, before making their first Canadian landfall at Iqaluit. The Labrador helicopters that the fifteen Cormorants were to replace had been flying for thirty-five years in a SAR role, and the British-Italian Cormorant represented a giant leap forward for the Canadian Forces.

With its three powerful engines, ice protection system, long-range capability, and large cargo area, the Cormorant was ideally suited for Canada's climate and terrain. Able to fly faster and higher than the Lab, over the next two years the Cormorants would be arriving at various bases. Comox, where 442 Squadron did the conversion training, was slated for five, CFB Greenwood in Nova Scotia four, and Gander, Newfoundland, and Trenton, Ontario, three each. The Halifax-based IMP Group Ltd., which had been performing overhauls on Auroras and Sea Kings for many years, had been contracted to maintain the Cormorants, and IMP technicians had accompanied the 19 Wing airmen on the delivery flights.

EH Industries had been formed in June 1980 by the British Westland and Italian Augusta helicopter companies to develop and market a replacement for the Sea Kings that both countries used. The maritime version of the EH101 was designed for anti-submarine, anti-ship surveillance and tracking, and search and rescue roles. The British especially wanted the EH 101(called the Merlin) to be able to operate off their Type 23 general-purpose frigates and asked that its physical dimensions to be limited by a frigate hangar. Westland was to build the front fuselage and main

Cormorant.

Courtesy of the DND

rotor blades, Augusta the rear fuselage, rotor head, and hydraulic system, and Fiat the gear box. The order for forty-four Merlin Mark Is was placed by the Royal Navy in 1985, with additional aircraft to replace the British Army Pumas and naval Sea Kings later. There was even a possibility that the EH101 could be chosen as a replacement for the U.S. presidential helicopter fleet.

Courtesy of the DND

Cormorant.

In Canada, the EH101 was a prime candidate to succeed the forty-year-old Sea Kings and the thirty-five-year-old Labradors in the Canadian Forces. After exhaustive and expensive studies, in 1992 Prime Minister Brian Mulroney signed a $4.8 billion contract with EHI to buy forty-three EH-101 helicopters. But this was an election year, and the purchase of the helicopters became a controversial political issue in the election campaign. Their price contributed to Prime Minister Kim Campbell's downfall, and, calling them the Cadillacs of the helicopter world, one of the first decisions that incoming Prime Minister Jean Chrétien took was to cancel the contract. While this campaign promise cost the government $500 million in penalties and much political embarrassment, there was worse to come. Replacement of the naval Sea Kings could be put off, but not the immediate need for SAR helicopters. When bids for the latter were solicited, the Westland-Augusta consortium mounted an aggressive Team Cormorant campaign, which proved that the EH101 was still the best value for the job. To the chagrin of the Prime Minister's Office, it found itself forced to condone the purchase of aircraft that it had refused earlier, or risk more penalties, legal and financial.

The fifteen Cormorants, a cheaper version of the EH101, were bought for $780 million, with Minister Eggleton saying at the October 30 acceptance ceremony, "The day I arrived the file was on my desk concerning the need to replace the Labrador. So, its great when you can come to an occasion like this and actually see it happening." Eggleton must have allowed himself a wry smile, for the Cormorant had been a political embarrassment for both his party and his boss. Rumour had it that, so as not to embarrass the Prime Minister, the decision to purchase replacements for the Sea Kings was to be put off until he left office, possibly in 2004. The Sea Kings would be flying until 2010.